ARRL's
TECH
Q&A

Your Quick & Easy Path to your FIRST Ham Radio License!

Seventh Edition

By Ward Silver, NØAX

Contributing Editor:
Mark Wilson, K1RO

Editorial Assistant
Maty Weinberg, KB1EIB

Production Staff:
David Pingree, N1NAS,
Senior Technical Illustrator

Jodi Morin, KA1JPA,
Assistant Production
Supervisor—Layout

Sue Fagan, KB1OKW,
Graphic Design Supervisor
—Cover Design

Michelle Bloom, WB1ENT,
Production Supervisor
—Layout

ARRL
the national association for
amateur radio®

This book may be used for Technician license exams given beginning July 1, 2018 and ending June 30, 2022. *QST* and the ARRL website (**www.arrl.org**) will have news about any rules changes affecting the Technician class license or any of the material in this book.

Feedback: We're interested in hearing your comments on this book and what you'd like to see in future editions. Please email comments to us at **pubsfdbk@ arrl.org**, including your name, call sign, email address, and the title, edition and printing of this book.

We strive to produce books without errors. Sometimes mistakes do occur, however. When we become aware of problems in our books (other than obvious typographical errors), we post corrections on the ARRL website. If you think you have found an error, please check **www.arrl.org** for corrections and supplemental material. If you don't find a correction there, please let us know by sending email to **pubsfdbk@arrl.org**.

CONTENTS

Welcome to the diverse group of individuals who make up amateur radio! There are nearly 750,000 amateurs, or "hams," in the United States alone and 3,000,000 around the world. Hams come from all walks of life, all ages, and every continent. Hams are busily communicating without regard to the geographic and political barriers that often separate humanity. This is the power of amateur radio — to communicate with each other directly, without any other commercial or government systems.

Hams come to amateur radio from many walks of life and many interests. Perhaps you intend to engage in public service for yourself and your community. Technical experimentation might be your interest, or you might be a habitual tinkerer who is always building, testing, using, and learning. Making new friends via the radio, keeping in touch as you travel, or exploring where a wireless signal can take you — these are all valuable and valued parts of the Amateur Service.

This book's Q&A format will help you review each question as you prepare to pass your ham radio licensing exam. This book's companion study guide, *The Ham Radio License Manual* (available at **arrl.org/shop**), explores most topics in more detail. There is additional material on ARRL's website, too — **www.arrl.org/ham-radio-license-manual**. You can also review and take practice exams at **www.arrl.org/examreview**. By reviewing this extra information, you will be better prepared to get on the air, have more fun, and be a more effective operator.

After you get your license, be sure to explore your local and online ham radio clubs and groups. This is where you'll find answers to your questions and help in getting on the air. You can start by asking the hams who administer your exam session — their organization can be your first "Elmer," or mentor. One of ham radio's strongest traditions is helping other hams learn and grow.

Most active radio amateurs in the United States are ARRL members. They realize that ARRL's training, sponsorship of activities, and representation both nationally and internationally are second to none. The book you're reading now, *ARRL's Tech Q&A*, is just one of many publications for all levels and interests in amateur radio. You don't need a license to join ARRL — just be interested in amateur radio and we are interested in you. It's as simple as that!

Ward Silver, NØAX
St. Charles, Missouri
March 2018

When to Expect New Books

A Question Pool Committee (QPC) consisting of representatives from the various Volunteer Examiner Coordinators (VECs) prepares the license question pools. The QPC establishes a schedule for revising and implementing new question pools. The current question pool revision schedule is as follows:

Question Pool	Current Study Guides	Valid Through
Technician (Element 2)	The ARRL Ham Radio License Manual, 4th Edition ARRL's Tech Q&A, 7th Edition	June 30, 2022
General (Element 3)	The ARRL General Class License Manual, 9th edition ARRL's General Q&A, 6th Edition	June 30, 2023
Amateur Extra (Element 4)	The ARRL Extra Class License Manual, 12th Edition ARRL's Extra Q&A, 5th Edition	June 30, 2024

As new question pools are released, ARRL will produce new study materials before the effective date of the new pools. Until then, the current question pools will remain in use, and current ARRL study materials, including this book, will help you prepare for your exam.

As the new question pool schedules are confirmed, the information will be published in *QST* and on the ARRL website at **www.arrl.org**.

Online Review and Practice Exams

Use this book with the online *ARRL Exam Review for Ham Radio* to take randomly generated practice exams using questions from the actual examination question pool. You won't have any surprises on exam day! Go to **www.arrl.org/examreview**.

About ARRL

The seed for amateur radio was planted in the 1890s, when Guglielmo Marconi began his experiments in wireless telegraphy. Soon he was joined by dozens, then hundreds, of others who were enthusiastic about sending and receiving messages through the air — some with a commercial interest, but others solely out of a love for this new communications medium. The United States government began licensing amateur radio operators in 1912.

By 1914, there were thousands of amateur radio operators — hams — in the United States. Hiram Percy Maxim, a leading Hartford, Connecticut inventor and industrialist, saw the need for an organization to unify this fledgling group of radio experimenters. In May 1914 he founded the American Radio Relay League (ARRL) to meet that need.

ARRL is the national association for Amateur Radio in the US. Today, with approximately 167,000 members, ARRL numbers within its ranks the vast majority of active radio amateurs in the nation and has a proud history of achievement as the standard-bearer in amateur affairs. ARRL's underpinnings as amateur radio's witness, partner, and forum are defined by five pillars: Public Service, Advocacy, Education, Technology, and Membership. ARRL is also International Secretariat for the International Amateur Radio Union, which is made up of similar societies in 150 countries around the world.

ARRL's Mission Statement: To advance the art, science, and enjoyment of amateur radio.

ARRL's Vision Statement: As the national association for Amateur Radio in the United States, ARRL:

• Supports the awareness and growth of amateur radio worldwide;
• Advocates for meaningful access to radio spectrum;
• Strives for every member to get involved, get active, and get on the air;
• Encourages radio experimentation and, through its members, advances radio technology and education;
     and
• Organizes and trains volunteers to serve their communities by providing public service and emergency communications.

At ARRL headquarters in the Hartford, Connecticut suburb of Newington, the staff helps serve the needs of members. ARRL publishes a monthly membership journal, *QST*; a bimonthly magazine for new hams, *On the Air*; a journal for experimenters, *QEX*; and a radiosport magazine *NCJ: National Contest Journal*, as well as newsletters and many publications covering all aspects of amateur radio. Its headquarters station, W1AW, transmits bulletins of interest to radio amateurs, and Morse code practice sessions. ARRL also coordinates an extensive field organization, which includes volunteers who provide technical information and other support services for radio amateurs as well as communications for public service activities. In addition, ARRL represents US radio amateurs to the Federal Communications Commission and other government agencies in the US and abroad.

Membership in ARRL means much more than receiving *QST* or *On the Air*. In addition to the services already described, ARRL offers membership services on a personal level, such as the Technical Information Service, where members can get answers — by phone, email, or the ARRL website — to all their technical and operating questions.

A bona fide interest in amateur radio is the only essential qualification of membership; an amateur radio license is not a prerequisite, although full voting membership is granted only to licensed radio amateurs in the US. Full ARRL membership gives you a voice in how the affairs of the organization are governed. ARRL policy is set by a Board of Directors (one from each of 15 Divisions). Each year, one-third of the ARRL Board of Directors stands for election by the full members they represent. The day-to-day operation of ARRL HQ is managed by a Chief Executive Officer and their staff.

Join ARRL Today! No matter what aspect of amateur radio attracts you, ARRL membership is relevant and important. There would be no amateur radio as we know it today were it not for ARRL. We would be happy to welcome you as a member! Join online at www.arrl.org/join. For more information about ARRL and answers to any questions you may have about amateur radio, write or call:

ARRL — The national association for Amateur Radio®

225 Main Street
Newington CT 06111-1400
Tel: 860-594-0200
FAX: 860-594-0259
email: hq@arrl.org
www.arrl.org

Prospective new radio amateurs call (toll-free):
800-32-NEW HAM (800-326-3942)
You can also contact ARRL via e-mail at **newham@arrl.org**
or check out the ARRL website at **www.arrl.org**

Get more from your Technician Class License with ARRL Membership

Membership in ARRL offers unique opportunities to advance and share your knowledge of amateur radio. For over 100 years, advancing the art, science, and enjoyment of amateur radio has been our mission. Your membership helps to ensure that new generations of hams continue to reap the benefits of the amateur radio community.

Here are just a few of the benefits you will receive with your annual membership. For a complete list visit, arrl.org/membership.

KNOWLEDGE

ARRL offers you a wealth of knowledge to advance your skills with lifelong learning courses, local clubs where you can meet and share ideas, and publications to help you keep up with the latest information from the world of ham radio.

ADVOCACY

ARRL is a strong national voice for preserving and protecting access to Amateur Radio Service frequencies.

SERVICES

From free FCC license renewals, to our Technical Information Service that answers calls and emails about your operating and technical concerns, ARRL offers a range of member services.

RESOURCES

Digital resources including email forwarding, product review archives, e-newsletters, and more.

PUBLICATIONS

Members receive digital access to all four ARRL monthly and bimonthly publications — *QST*, the membership journal of ARRL; *On the Air*, an introduction to the world of amateur radio; *QEX*, which covers topics related to amateur radio and radio communications experimentation, and *National Contest Journal* (*NCJ*), covering radio contesting.

Two Easy Ways to Join

CALL Member Services toll free at
1-888-277-5289

ONLINE Go to our secure website at
arrl.org/join

The Technician License

Earning a Technician amateur radio license begins your enjoyment of ham radio. Topics covered by the exam provide you with a good introduction to basic radio, and there is no difficult math or electronics background required. You are sure to find the operating privileges available to a Technician licensee to be worth the time spent learning about amateur radio. After passing the exam, you will be able to operate on every frequency above 50 megahertz available to the Amateur Service. You also gain privileges on the traditional "high-frequency (HF)" 80, 40, 15, and 10 meter amateur bands. With this broad set of operating privileges, you'll be ready to experience the excitement of amateur radio!

Perhaps your interest is in amateur radio's long history of public service, such as providing emergency communications in time of need. You might have experience with computer networks leading you to explore the digital mode technology used in ham radio. If your eyes turn to the stars on a clear night, you might enjoy tracking the amateur satellites and using them to relay your signals to other amateurs around the world! Your whole family can enjoy amateur radio, taking part in outdoor activities such as ARRL Field Day and mobile operating during a vacation or weekend drive.

An Overview of Amateur Radio

Earning an amateur radio license is a special achievement. The more than 750,000 people in the US who call themselves amateur radio operators, or hams, are part of a global family. Radio signals do not know territorial boundaries, so hams have a unique ability to enhance international goodwill. Hams become ambassadors of their country every time they put their stations on the air.

Radio amateurs provide a voluntary, noncommercial, communication service, without any type of payment except the personal satisfaction they feel from a job well done! This is especially true during natural disasters or other emergencies when the normal lines of communication are out of service. In the aftermath of the devasting hurricanes of 2017, for example, hams responded by traveling to Caribbean islands, southeast Texas, and Florida's western cities. They set up temporary communication networks, transferred information into and out of the affected areas, and provided communication support for the authorities until normal systems were working again. They were supported by thousands of hams around the US and the rest of the world. In every US county and city, organized groups of amateur operators train and prepare to support their communities during disasters and emergencies of every type.

Hams have made many important contributions to the field of electronics and communications since amateur radio's beginnings a century ago and this tradition continues today. Today, hams relay signals through their own satellites, bounce signals off the moon, relay messages automatically through computerized radio networks and use any number of other "exotic" communications techniques.

Amateur radio experimentation is yet another reason many people become part of this self-trained group of trained operators, technicians and electronics experts — an asset to any country. Amateurs talk from hand-held transceivers through mountaintop repeater stations that relay their signals to other hams' cars or homes or through the Internet around the world. Hams establish wireless data networks, send their own television signals, and talk with other hams around the world by voice. Keeping alive a distinctive traditional skill, they also tap out messages in Morse code.

Who Can Be a Ham...and How?

The US government, through the Federal Communications Commission (FCC), grants all US amateur radio licenses. Because of amateur radio's tremendously flexible and self-organizing nature, amateurs are expected to know more about their equipment and operating techniques. Unlike other radio services, amateurs organize their own methods of communication, are encouraged to build and repair their own equipment, perform experiments with antennas and with radio propagation, and invent their own protocols and networks. The FCC licensing process ensures that amateurs have the necessary operating skill and electronics know-how to use that flexibility wisely and not interfere with other radio services.

It doesn't matter how old you are or whether you're a US citizen. If you pass the examination, the Commission will issue you an amateur license. Any person (except the agent of a foreign government) may take the exam and, if successful, receive an amateur license. It's important to understand that if a citizen of a foreign country receives an amateur license in this manner, he or she is a US Amateur Radio operator. (This should not be confused with a reciprocal permit, which is covered in the questions of subelement T1.)

License Structure

Anyone earning a new amateur radio license can earn one of three license classes — Technician, General and Amateur Extra. Higher-class licenses have more comprehensive examinations. In return for passing a more difficult exam you earn more frequency privileges (frequency space in the radio spectrum). The vast majority of beginners earn the most basic license, the Technician, before beginning to study for the other licenses.

Table 1 lists the amateur license classes you can earn, along with a brief description of their exam requirements and operating privileges. A Technician license gives you the freedom to develop operating and technical skills through on-the-air experience. These skills will help you upgrade to a higher class of license and obtain additional privileges.

The Technician exam, called Element 2, covers some basic radio fundamentals and knowledge of some of the rules and regulations in Part 97 of the FCC Rules. With a little study you'll soon be ready to pass the exam.

Each step up the amateur radio license ladder requires the applicant to have passed the lower exams. So if you want to start out as a General class or even an Amateur Extra class licensee, you must first have passed the Technician written exam. You retain credit for all the exam elements of any license class you hold. For example, if you hold a Technician license, you will only have to pass the Element 3 General class written exam to obtain a General class license.

Although there are also other amateur license classes, the FCC is no longer

Table 1
Amateur Operator Licenses

License Class	Written Exam	Privileges
Technician	Basic theory and regulations (Element 2)	All above 50 MHz and limited HF privileges
General	Basic theory and regulations; General theory and regulations (Elements 2 and 3)	All except those reserved for Advanced and Amateur Extra
Amateur Extra	All lower exam elements, plus Amateur Extra theory (Elements 2, 3 and 4)	All amateur privileges

issuing new licenses for them. The Novice license was long considered the beginner's license. Exams for this license were discontinued as of April 15, 2000. The FCC also stopped issuing new Advanced class licenses on that date. They will continue to renew previously issued licenses, however, so you will probably meet some Novice and Advanced class licensees on the air.

As a Technician, you can use a wide range of frequency bands — all amateur bands above 50 MHz (megahertz), in fact. (See **Table 2** and **Figure 1**.) You'll be able to use point-to-point or repeater communications on VHF, use packet radio and other digital modes and networks, even access orbiting satellites or bounce a signal off meteor trails! You can use your operating skills to provide public service through emergency communications and message handling.

Station Call Signs

Many years ago, by international agreement, the nations of the world decided to allocate certain call sign prefixes to each country. This means that if you hear a radio station call sign beginning with W or K, for example, you know the station is licensed by the United States. A call sign beginning with the letter G is licensed by Great Britain, and a call sign beginning with VE is from Canada. (All of the amateur call sign prefixes are listed in a table on ARRL's website, **www.arrl.org**.)

The International Telecommunication Union (ITU) radio regulations outline the basic principles used in forming amateur call signs. According to these regulations, an amateur call sign must be made up of one or two characters (the first one may be a numeral) as a prefix, followed by a numeral, and then a suffix of up to three letters. The prefixes W, K, N and A are used in the United States. When the letter A is used in a US amateur call sign, it will always be with a two-letter prefix, AA to AL. The continental US is divided into 10 amateur radio call districts (commonly called "call areas"), numbered 0 through 9. **Figure 2** is a map showing the US call districts.

You may keep the same call sign when you change license class, if you wish. You must indicate that you want to receive a new call sign when you apply for the exam or change your address.

Table 2

US Amateur Bands

Amateurs wishing to operate on either 2,200 or 630 meters must first register with the Utilities Technology Council online at **https://utc.org/plc-database-amateur-notification-process/**. You need only register once for each band.

2,200 Meters (135 kHz)

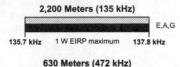

135.7 kHz   1 W EIRP maximum   137.8 kHz   E,A,G

630 Meters (472 kHz)

472 kHz   479 kHz   E,A,G
5 W EIRP maximum, except in Alaska within 496 miles of Russia where the power limit is 1 W EIRP.

160 Meters (1.8 MHz)
Avoid interference to radiolocation operations from 1.900 to 2.000 MHz

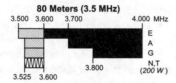

1.800   1.900   2.000 MHz   E,A,G

80 Meters (3.5 MHz)

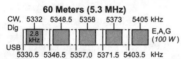

3.500 3.600 3.700   4.000 MHz
E
A
G
N,T
(200 W)
3.525 3.600   3.800

60 Meters (5.3 MHz)

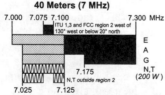

CW, Dig   5332 5348.5 5358 5373 5405 kHz
2.8 kHz
USB   5330.5 5346.5 5357.0 5371.5 5403.5 kHz
E,A,G (100 W)

General, Advanced, and Amateur Extra licensees may operate on these five channels on a secondary basis with a maximum effective radiated power (ERP) of 100 W PEP relative to a half-wave dipole. Permitted operating modes include upper sideband voice (USB), CW, RTTY, PSK31 and other digital modes such as PACTOR III. Only one signal at a time is permitted on any channel.

40 Meters (7 MHz)

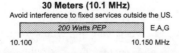

7.000 7.075 7.100   7.300 MHz
ITU 1,3 and FCC region 2 west of 130° west or below 20° north
E
A
G
N,T
(200 W)
7.175
N,T outside region 2
7.025   7.125

See Sections 97.305(c), 97.307(f)(11) and 97.301(e). These exemptions do not apply to stations in the continental US.

30 Meters (10.1 MHz)
Avoid interference to fixed services outside the US.

200 Watts PEP   E,A,G

10.100   10.150 MHz

20 Meters (14 MHz)

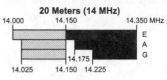

14.000   14.150   14.350 MHz
E
A
G
14.175
14.025   14.150   14.225

17 Meters (18 MHz)

E,A,G

18.068   18.110   18.168 MHz

15 Meters (21 MHz)

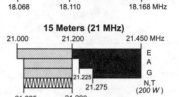

21.000   21.200   21.450 MHz
E
A
G
N,T (200 W)
21.225
21.275
21.025   21.200

12 Meters (24 MHz)

E,A,G

24.890   24.930   24.990 MHz

10 Meters (28 MHz)

28.000 28.300   29.700 MHz
E,A,G
N,T (200 W)
28.000   28.500

6 Meters (50 MHz)

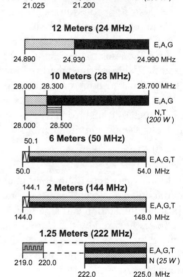

50.1
E,A,G,T
50.0   54.0 MHz

2 Meters (144 MHz)

144.1
E,A,G,T
144.0   148.0 MHz

1.25 Meters (222 MHz)

219.0 220.0
E,A,G,T
N (25 W)
222.0   225.0 MHz

*Geographical and power restrictions may apply to all bands above 420 MHz. See FCC Part 97.303 for information about your area.

70 cm (420 MHz)*

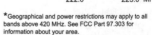

E,A,G,T
420.0   450.0 MHz

33 cm (902 MHz)*

E,A,G,T
902.0   928.0 MHz

23 cm (1240 MHz)*

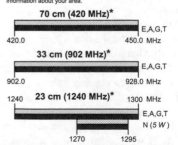

1240   1300 MHz
E,A,G,T
N (5 W)
1270   1295

US AMATEUR POWER LIMITS

FCC 97.313 An amateur station must use the minimum transmitter power necessary to carry out the desired communications. **(b)** No station may transmit with a transmitter power exceeding **1.5 kW PEP.**

All licensees except Novices are authorized all modes on the following frequencies:

2300-2310	MHz	
2390-2450	MHz	
3300-3500	MHz	
5650-5925	MHz	
10.0-10.5	GHz*	
24.0-24.25	GHz	

47.0-47.2	GHz
76.0-81.0	GHz
122.25-123.0	GHz
134-141	GHz
241-250	GHz
All above 275 GHz	

* No pulse emissions

```
────────KEY────────
Note:
CW operation is permitted throughout all
amateur bands.

MCW is authorized above 50.1 MHz,
except for 144.0-144.1 and 219-220 MHz.

Test transmissions are authorized above
51 MHz, except for 219-220 MHz

▒▒▒ = RTTY and data
▓▓▓ = phone and image
〰〰 = CW only
≡≡ = SSB phone
▓▓▓ = USB phone, CW, RTTY
          and data.
ꟿꟿ = Fixed digital message
          forwarding systems only

E = Amateur Extra
A = Advanced
G = General
T = Technician
N = Novice
```

See *www.arrl.org* for more detailed band plans.

The FCC also has a vanity call sign system that allows you to change your call sign from the one the FCC assigned you, to one that you like better — provided the call sign is available for use. While there is no fee for an amateur radio license, there is a fee for the selection of a vanity call sign. The current fee and details of the vanity call sign system are available on the ARRL website at **www.arrl. org/vanity-call-signs**.

How to Use This Book

The Element 2 exam consists of 35 questions taken from a pool of around 450 questions. The *ARRL's Tech Q&A* is designed to help you learn about every question in the Technician exam question pool. Every question is presented just as it is in the question pool and as you will encounter it on the exam. Following every question is a short explanation of the answer.

Each chapter of the book covers one subelement from the question pool, beginning with the FCC Rules and ending with Safety. You may study the questions from beginning to end or select topics in an order that appeals to you.

If you are new to radio, you will probably find it easier to begin with the questions in subelement T3, T4 and T7 to learn about amateur equipment and the basics of radio signals. Then you can move on to the more technical topics covered by T5, T6, T8 and T9. Once you've learned about how radios work, the subelements on operating (T2) and FCC Rules (T1) will make more sense. Finish up with T0 — Safety — and you'll be ready for your exam!

The *ARRL Ham Radio License Manual* is a good reference companion to the *Tech Q&A*. At the end of the

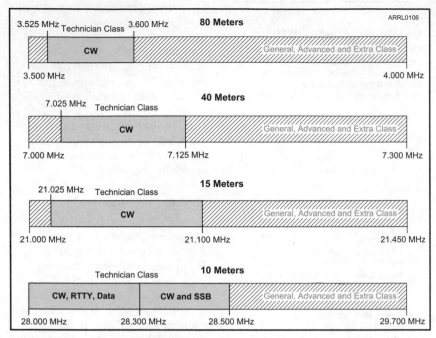

Figure 1 — This chart details the HF privileges available to Technician licensees.

explanation for every question, there is a reference to the page in the *Ham Radio License Manual* where you can find a discussion of the topics associated with the question.

There is additional material at the ARRL's website **www.arrl.org/ham-radio-license-manual** if you need extra help. In particular, there are links to math tutorials and every math problem on the exam is completely worked out to show you how it's done. To make the best use of the online reference material, bookmark the *Ham Radio License Manual* website to use as an online reference while you study.

The ARRL's New Ham Desk can answer questions emailed to **newham@arrl.org**. Your question may be answered directly or you might be directed to more instruction material. The New Ham Desk can also help you find a local ham to answer questions. Studying with a friend makes learning the material more fun as you help each other over the rough spots and you'll have someone to celebrate with after passing the exam!

Earning a License

All US amateur exams are administered by Volunteer Examiners who are certified by a Volunteer Examiner Coordinator (VEC) that processes the examination paperwork and license applications for the FCC. A Question Pool Committee selected by the Volunteer Examiner Coordinators maintains the question pools for all amateur exams.

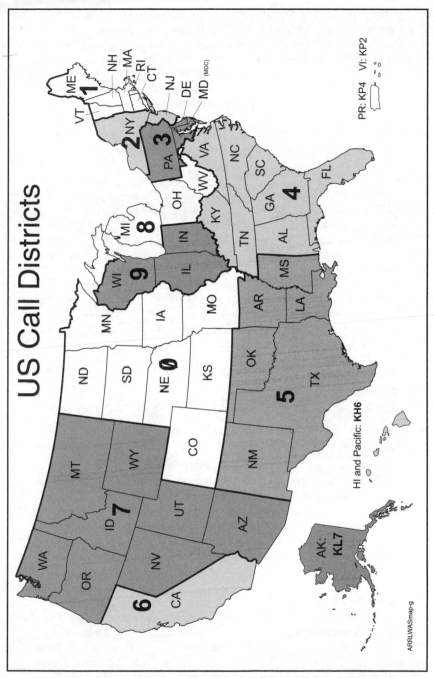

US Call Districts

PR: KP4   VI: KP2

HI and Pacific: **KH6**

AK: **KL7**

ARRLWASmap-g

Figure 2—There are 10 US call districts or areas. Hawaii and all Pacific possessions are part of the sixth call area and Alaska is part of the seventh. Puerto Rico and the US Virgin Islands are part of the fourth district.

Once you make the commitment to study and learn what it takes to pass the exam, you will accomplish your goal. Many people pass the exam on their first try, so if you study the material and are prepared, chances are good that you will soon have your license. It may take you more than one attempt to pass the Technician license exam, but that's okay. There is no limit to how many times you can take it. Many Volunteer Examiner teams have several exam versions available, so you may even be able to try the exam again at the same exam session. Time and available exam versions may limit the number of times you can try the exam at a single exam session. If you don't pass after a couple of tries you will certainly benefit from more study of the question pools before you try again.

License Examinations

The FCC allows Volunteer Examiners to select the questions for an amateur exam, but they must use the questions exactly as they are released by the VEC that coordinates the test session. If you attend a test session coordinated by the ARRL/VEC, your test will be designed by the ARRL/VEC or by a computer program created by the VEC. The questions and answers will be exactly as they are printed in this book.

Before you can take an FCC exam, you'll have to fill out a copy of the National Conference of Volunteer Examiner Coordinators (NCVEC) Quick Form 605. This form is used as an application for a new license or an upgraded license. The NCVEC Quick Form 605 is only used at license exam sessions. This form includes some information that the Volunteer Examiner Coordinator's office will need to process your application with the FCC. See **Figure 3**.

You should not use an NCVEC Quick Form 605 to apply for a license renewal or modification with the FCC. Never mail these forms to the FCC, because that will result in a rejection of the application. Likewise, an FCC Form 605 can't be used for a license exam application.

Finding an Exam Session

You can locate upcoming exam sessions in your area by using ARRL's online Exam Search page. Browse to **www.arrl.org**, and click the "Licensing, Education & Training" button to find complete information about taking a licensing exam. Registration deadlines and the time and location of the exams, are mentioned prominently in publicity releases about upcoming sessions. You can also contact the ARRL/VEC office directly or watch for announcements in *QST* or on the ARRL website. Many local clubs sponsor exams, so they are another good source of information on exam opportunities.

Taking the Exam

By the time examination day rolls around, you should have already prepared yourself. This means getting your schedule, supplies and mental attitude ready. Plan your schedule so you'll get to the examination site with plenty of time to spare. There's no harm in being early. In fact, you might have time to meet and talk with another applicant which is a great way to calm pretest nerves.

What supplies will you need? While this is probably your first license, if you have any current amateur licenses or a Certificate of Successful Completion of

Examination (CSCE), bring the original and a photocopy. Bring along several sharpened number 2 pencils and two pens (blue or black ink). Be sure to have a good eraser. A pocket calculator may also come in handy. You may use a programmable calculator if that is the kind you have, but take it into your exam "empty" (cleared of all programs and constants in memory). Don't program equations ahead of time, because you may be asked to demonstrate that there is nothing in the calculator memory. The examining team has the right to refuse a candidate the use of any calculator that they feel may contain information for the test or could otherwise be used to cheat on the exam.

The Volunteer Examiner team is required to check two forms of identifica-

NCVEC QUICK-FORM 605 APPLICATION
AMATEUR OPERATOR/PRIMARY STATION LICENSE

SECTION 1 - TO BE COMPLETED BY APPLICANT

PRINT LAST NAME	SUFFIX (Jr., Sr.)	FIRST NAME	M.I.	STATION CALL SIGN (IF LICENSED)
SOMMA		MARIA		KB1KJC

MAILING ADDRESS (Number and Street or P.O. Box): 225 MAIN ST.

FEDERAL REGISTRATION NUMBER (FRN) - IF NONE, THEN SOCIAL SECURITY NUMBER (SSN): 0009876543

CITY: NEWINGTON   STATE CODE: CT   ZIP CODE (5 or 9 Numbers): 06111

DAYTIME TELEPHONE NUMBER (Include Area Code):

E-MAIL ADDRESS (MANDATORY TO RECEIVE LICENSE NOTIFICATION EMAIL FROM FCC): KB1KJC @ arrl.org

Basic Qualification Question:
Has the Applicant or any party to this application, or any party directly or indirectly controlling the Applicant, ever been convicted of a felony by any state or federal court? ☐ YES ☒ NO
If "YES", see "FCC BASIC QUALIFICATION QUESTION INSTRUCTIONS AND PROCEDURES" on the back of this form.

I HEREBY APPLY FOR (Make an X in the appropriate box(es)):

☐ EXAMINATION for a new license grant

☒ EXAMINATION for upgrade of my license class

☐ CHANGE my name on my license to my new name

Former Name: _____
(Last name) (Suffix)   (First name)   (MI)

☐ CHANGE my mailing address to **above** address

☐ CHANGE my station **call sign** systematically
Applicant's Initials: To Confirm _____

☐ RENEWAL of my license grant
Exp. Date: _____

Do you have another license application on file with the FCC which has not been acted upon?	PURPOSE OF OTHER APPLICATION	PENDING FILE NUMBER (FOR VEC USE ONLY)

I certify that:
- I waive any claim to the use of any particular frequency regardless of prior use by license or otherwise;
- All statements and attachments are true, complete and correct to the best of my knowledge and belief and are made in good faith;
- I am not a representative of a foreign government;
- I am not subject to a denial of Federal benefits pursuant to Section 5301 of the Anti-Drug Abuse Act of 1988, 21 U.S.C. § 862;
- The construction of my station will NOT be an action which is likely to have a significant environmental effect (See 47 CFR Sections 1.1301-1.1319 and Section 97.13(a));
- I have read and WILL COMPLY with Section 97.13(c) of the Commission's Rules regarding RADIOFREQUENCY (RF) RADIATION SAFETY and the amateur service section of OST/OET Bulletin Number 65.

Signature of Applicant:
X *Maria Somma*   Date Signed: 03 - 13 - 2018

SECTION 2 - TO BE COMPLETED BY ALL ADMINISTERING VEs

Applicant is qualified for operator license class:

☐ NO NEW LICENSE OR UPGRADE WAS EARNED

☐ TECHNICIAN   Element 2

☒ GENERAL   Elements 2 and 3

☐ AMATEUR EXTRA   Elements 2, 3 and 4

DATE OF EXAMINATION SESSION: 03 - 13 - 2018
EXAMINATION SESSION LOCATION: NEWINGTON CT
VEC ORGANIZATION: ARRL
VEC RECEIPT DATE:

I CERTIFY THAT I HAVE COMPLIED WITH THE ADMINISTERING VE REQUIRMENTS IN PART 97 OF THE COMMISSION'S RULES AND WITH THE INSTRUCTIONS PROVIDED BY THE COORDINATING VEC AND THE FCC.

	VEs STATION CALL SIGN	VEs SIGNATURE (Must match name)	DATE SIGNED
1st VEs NAME (Print First, MI, Last, Suffix): Jenny Harts	NN1AG	*Jenny Harts*	3/13/18
2nd VEs NAME (Print First, MI, Last, Suffix): Rose-Ann Lawrence	KB1DMW	*Rose-Ann Lawrence*	03-13-18
3rd VEs NAME (Print First, MI, Last, Suffix): Perry Green	WY1O	*Perry Green*	03/13/18

DO NOT SEND THIS FORM TO FCC - THIS IS NOT AN FCC FORM.
IF THIS FORM IS SENT TO FCC, FCC WILL RETURN IT TO YOU WITHOUT ACTION.

NCVEC FORM 605 - September 2017
FOR VE/VEC USE ONLY - Page 1

Figure 3—At the test session, the Volunteer Examiners will help you fill out an NCVEC Quick Form 605, which will be filed with the FCC.

tion before you enter the test room. You can use a driver's license, a piece of mail addressed to you, a birth certificate, or a photo ID of some type. If you have an original amateur radio license, bring it — not a photocopy.

The following description of the testing procedure applies to exams coordinated by the ARRL/VEC, although many other VECs use a similar procedure.

Written Test

The examiner will give each applicant a test booklet, an answer sheet and scratch paper. You'll be shown where to sign your name and after that, you're on your own. The first thing to do is read the instructions.

Next, check the examination to see that all pages and questions are there. If not, report this to the examiner immediately. When filling in your answer sheet make sure your answers are marked next to the numbers that correspond to each question.

Go through the entire exam, and answer the easy questions first. Next, go back to the beginning and try the harder questions. Leave the really tough questions for last.

If you don't know the answer to a question, make your best guess. There is no additional penalty for answering incorrectly. If you have to guess, do it intelligently: At first glance, you may find that you can eliminate one or more "distractors." Of the remaining responses, more than one may seem correct; only one is the best answer, however. To the applicant who is fully prepared, incorrect distractors to each question are obvious. Nothing beats preparation!

After you've finished, check the examination thoroughly. You may have read a question wrong or goofed in your arithmetic. Don't be overconfident. There's no rush, so take your time. Think and check your answer sheet. When you feel you've done your best, return the test booklet, answer sheet and scratch pad to the examiner.

The Volunteer Examiner team will grade the exam while you wait. The passing grade is 74%. (That means 26 out of 35 questions correct with up to 9 incorrect answers on the Element 2 exam.) You will receive a Certificate of Successful Completion of Examination (CSCE — see **Figure 4**) showing all exam elements that you pass at that exam session. That certificate is valid for 365 days. Use it as proof that you passed those exam elements so you won't have to take them over again next time you take a license exam.

Forms and Procedures

To renew or modify a license, you can file a copy of FCC Form 605. In addition, hams who have held a valid license that has expired within the past two years may apply for reinstatement with an FCC Form 605.

Licenses are normally good for 10 years. Your application for a license renewal must be submitted to the FCC no more than 90 days before the license expires. (We recommend you submit the application for renewal between 90 and 60 days before your license expires.) If the FCC receives your renewal application before the license expires, you may continue to operate until your new license is granted and shows up in the FCC database, even if it is past the expiration date.

If you forget to apply before your license expires, you may still be able to

American Radio Relay League VEC
Certificate of Successful Completion of Examination ARRL *The national association for* AMATEUR RADIO®

NOTE TO VE TEAM:
COMPLETELY CROSS OUT ALL BOXES BELOW THAT DO NOT APPLY TO THIS CANDIDATE.

Test Site
(City/State): NEWINGTON, CT   Test Date: 03-13-18

The applicant named herein has presented valid proof for the exam element credit(s) indicated below.

Element 3 credit
Element 4 credit

CREDIT for ELEMENTS PASSED VALID FOR 365 DAYS
You have passed the written element(s) indicated at right. Your will be given credit for the appropriate examination element(s), for up to 365 days from the date shown at the top of this certificate.

EXAM ELEMENTS EARNED

LICENSE UPGRADE NOTICE
If you also hold a valid FCC-issued Amateur radio license grant, this Certificate validates temporary operation with the operating privileges of your new operator class (see Section 97.9[b] of the FCC's Rules) until you are granted the license for your new operator class, or for a period of 365 days from the test date stated above on this certificate, whichever comes first.

Passed written Element 3

NEW LICENSE CLASS EARNED

LICENSE STATUS INQUIRIES
You can find out if a new license or upgrade has been "granted" by the FCC. For on-line inquiries see the FCC Web at http://wireless.fcc.gov/uls/ ("Click on Search Licenses" button), or see the ARRL Web at http://www.arrl.org/fcc/search; or by calling FCC toll free at 888-225-5322; or by calling the ARRL at 1-860-594-0300 during business hours. **Allow 15 days from the test date before calling.**

TECHNICIAN
GENERAL
EXTRA
NONE

THIS CERTIFICATE IS NOT A LICENSE, PERMIT, OR ANY OTHER KIND OF OPERATING AUTHORITY IN AND OF ITSELF. THE ELEMENT CREDITS AND/OR OPERATING PRIVILEGES THAT MAY BE INDICATED IN THE LICENSE UPGRADE NOTICE ARE VALID FOR 365 DAYS FROM THE TEST DATE. THE HOLDER NAMED HEREON MUST ALSO HAVE BEEN GRANTED AN AMATEUR RADIO LICENSE ISSUED BY THE FCC TO OPERATE ON THE AIR.

Candidate's Signature Maria Somma Call Sign KB1KJC (If none, write none)

Candidate's Name MARIA SOMMA

Address 225 MAIN ST.

City NEWINGTON State CT ZIP 06111

VE #1 Penny Harte Call Sign N1NAG
Signature

VE #2 Rose Ana Lawrence Call Sign KB1DMW
Signature

VE #3 Perry Green Call Sign WY1O
Signature

COPIES: WHITE-Candidate, YELLOW-VE Team, PINK-ARRL VEC
MVE 07/2015

Figure 4 — The CSCE (Certificate of Successful Completion of Examination) is your test session receipt that serves as proof that you have completed one or more exam elements. It can be used at other test sessions for 365 days.

renew your license without taking another exam. There is a two-year grace period, during which you may apply for renewal of your expired license. Use an FCC Form 605 to apply for reinstatement (and your old call sign). If you apply for reinstatement of your expired license under this two-year grace period, you may not operate your station until your new license is issued.

If you move or change addresses, you should use an FCC Form 605 to notify the FCC of the change. If your license is lost or destroyed, however, just write a letter to the FCC explaining why you are requesting a new copy of your license.

You can ask one of the Volunteer Examiner Coordinators' offices to file your renewal application electronically if you don't want to mail the form to the FCC. You must still mail the form to the VEC, however. The ARRL/VEC office will electronically file application forms. This service is free for any ARRL member.

Electronic Filing

You can also file your license renewal or address modification using the FCC's Universal Licensing System (ULS) website, **www.fcc.gov/uls**. To use ULS, you must have an FCC Registration Number, or FRN. Obtain your FRN by registering with the Commission Registration System, known as CORES.

Described as an agency-wide registration system for anyone filing applications with or making payments to the FCC, CORES will assign a unique 10-digit FCC Registration Number (FRN) to all registrants. All Commission systems that handle financial, authorization of service, and enforcement activities will use the FRN. The FCC says use of the FRN will allow it to more rapidly verify fee payment. Amateurs mailing payments to the FCC — for example as part of a vanity call sign application — would include their FRN on FCC Form 159.

The online filing system and further information about CORES is available by visiting the FCC web home page, **www.fcc.gov**, and clicking on the Commission Registration System link. Follow the directions on the website. It is also possible to register on CORES using a paper Form 160.

When you register with CORES you must supply a Taxpayer Identification Number, or TIN. For individuals, this is usually a Social Security Number. Club stations that do not have an EIN register as exempt. Anyone can register on CORES and obtain an FRN. You don't need a license to be registered.

Once you have registered on CORES and obtained your FRN, you can proceed to renew or modify your license using the Universal Licensing System by clicking on the "Online Filing" button. Follow the directions provided on the web page to connect to the FCC's ULS database.

Paper Filing

If you decide to "do the paperwork" on real paper instead of online, you'll need to get a blank FCC 605 Main Form. This is not difficult! You can get FCC 605 Main Form with detailed instructions from the FCC Forms web page — **www.fcc.gov/licensing-databases/forms.**

The ARRL VEC offers an overview of the various FCC and VEC forms and their uses online at **www.arrl.org/fcc-forms.**

And Now, Let's Begin

The complete Technician question pool (Element 2) is printed in this book. Each chapter lists all the questions for a particular subelement (such as Operating Procedures — T2). A brief explanation about the correct answer is given after each question.

Table 3 shows the study guide or syllabus for the Element 2 exam as released by the Volunteer Examiner Coordinators' Question Pool Committee in January 2018. The syllabus lists the topics to be covered by the Technician exam, and so forms the basic outline for the remainder of this book. Use the syllabus to guide your study.

The question numbers used in the question pool refer to this syllabus. Each question number begins with a syllabus point number (for example, T0C or T1A). The question numbers end with a two-digit number. For example, question T3B09 is the ninth question about the T3B syllabus topics.

The Question Pool Committee designed the syllabus and question pool so there are the same number of topics in each subelement as there are exam questions from that subelement. For example, three exam questions on the Technician exam must be from the "Operating Procedures" subelement, so there are three groups for that topic. These are numbered T2A, T2B, and T2C. While not a requirement of the FCC Rules, the Question Pool Committee recommends that one question be taken from each group to make the best possible license exams.

Good luck with your studies!

Table 3

Technician Class Syllabus

Effective July 1, 2018 to June 30, 2022

SUBELEMENT T1 — FCC Rules, descriptions, and definitions for the Amateur Radio Service, operator and station license responsibilities

[6 Exam Questions — 6 Groups]

T1A   Amateur Radio Service: purpose and permissible use of the Amateur Radio Service, operator/primary station license grant; Meanings of basic terms used in FCC rules; Interference; RACES rules; Phonetics; Frequency Coordinator

T1B   Authorized frequencies: frequency allocations; ITU; emission modes; restricted sub-bands; spectrum sharing; transmissions near band edges; contacting the International Space Station; power output

T1C   Operator licensing: operator classes; sequential and vanity call sign systems; international communications; reciprocal operation; places where the Amateur Radio Service is regulated by the FCC; name and address on FCC license database; license term; renewal; grace period

T1D   Authorized and prohibited transmission: communications with other countries; music; exchange of information with other services; indecent language; compensation for use of station; retransmission of other amateur signals; codes and ciphers; sale of equipment; unidentified transmissions; one-way transmission

T1E   Control operator and control types: control operator required; eligibility; designation of control operator; privileges and duties; control point; local, automatic and remote control; location of control operator

T1F   Station identification; repeaters; third-party communications; club stations; FCC inspection

SUBELEMENT T2 — Operating Procedures

[3 Exam Questions — 3 Groups]

T2A   Station operation: choosing an operating frequency; calling another station; test transmissions; procedural signs; use of minimum power; choosing an operating frequency; band plans; calling frequencies; repeater offsets

T2B   VHF/UHF operating practices: SSB phone; FM repeater; simplex; splits and shifts; CTCSS; DTMF; tone squelch; carrier squelch; phonetics; operational problem resolution; Q signals

T2C   Public service: emergency and non-emergency operations; applicability of FCC rules; RACES and ARES; net and traffic procedures; operating restrictions during emergencies

SUBELEMENT T3 — Radio wave characteristics: properties of radio waves; propagation modes

[3 Exam Questions — 3 Groups]

T3A   Radio wave characteristics: how a radio signal travels; fading; multipath; polarization; wavelength vs absorption; antenna orientation

T3B   Radio and electromagnetic wave properties: the electromagnetic spectrum; wavelength vs frequency; nature and velocity of electromagnetic waves; definition of UHF, VHF, HF bands; calculating wavelength

T3C   Propagation modes: line of sight; sporadic E; meteor and auroral scatter and reflections; tropospheric ducting; F layer skip; radio horizon

SUBELEMENT T4 — Amateur radio practices and station set-up

[2 Exam Questions — 2 Groups]

T4A   Station setup: connecting microphones; reducing unwanted emissions; power source; connecting a computer; RF grounding; connecting digital equipment; connecting an SWR meter

T4B   Operating controls: tuning; use of filters; squelch function; AGC; transceiver operation; memory channels

SUBELEMENT T5 — Electrical principles: math for electronics; electronic principles; Ohm's Law

[4 Exam Questions — 4 Groups]

T5A   Electrical principles, units, and terms: current and voltage; conductors and insulators; alternating and direct current; series and parallel circuits

T5B   Math for electronics: conversion of electrical units; decibels; the metric system

T5C   Electronic principles: capacitance; inductance; current flow in circuits; alternating current; definition of RF; definition of polarity; DC power calculations; impedance

T5D   Ohm's Law: formulas and usage; components in series and parallel

SUBELEMENT T6 — Electrical components; circuit diagrams; component functions

[4 Exam Questions — 4 Groups]

T6A   Electrical components: fixed and variable resistors; capacitors and inductors; fuses; switches; batteries

T6B   Semiconductors: basic principles and applications of solid state devices; diodes and transistors

T6C   Circuit diagrams; schematic symbols

T6D   Component functions: rectification; switches; indicators; power supply components; resonant circuit; shielding; power transformers; integrated circuits

SUBELEMENT T7 — Station equipment: common transmitter and receiver problems; antenna measurements; troubleshooting; basic repair and testing

[4 Exam Questions — 4 Groups]

T7A   Station equipment: receivers; transmitters; transceivers; modulation; transverters; transmit and receive amplifiers

T7B   Common transmitter and receiver problems: symptoms of overload and overdrive; distortion; causes of interference; interference and consumer electronics; part 15 devices; over-modulation; RF feedback; off frequency signals

T7C   Antenna measurements and troubleshooting: measuring SWR; dummy loads; coaxial cables; causes of feed line failures

T7D   Basic repair and testing: soldering; using basic test instruments; connecting a voltmeter, ammeter, or ohmmeter

SUBELEMENT T8 — Modulation modes: amateur satellite operation; operating activities; non-voice and digital communications

[4 Exam Questions — 4 Groups]

T8A   Modulation modes: bandwidth of various signals; choice of emission type

T8B   Amateur satellite operation; Doppler shift; basic orbits; operating protocols; transmitter power considerations; telemetry and telecommand; satellite tracking

T8C   Operating activities: radio direction finding; radio control; contests; linking over the internet; grid locators

T8D   Non-voice and digital communications: image signals; digital modes; CW; packet radio; PSK31; APRS; error detection and correction; NTSC; amateur radio networking; Digital Mobile/Migration Radio

SUBELEMENT T9 — Antennas and feed lines

[2 Exam Questions — 2 Groups]

T9A   Antennas: vertical and horizontal polarization; concept of gain; common portable and mobile antennas; relationships between resonant length and frequency; concept of dipole antennas

T9B   Feed lines: types, attenuation vs frequency, selecting; SWR concepts; Antenna tuners (couplers); RF Connectors: selecting, weather protection

SUBELEMENT T0 — Electrical safety: AC and DC power circuits; antenna installation; RF hazards

[3 Exam Questions — 3 Groups]

T0A   Power circuits and hazards: hazardous voltages; fuses and circuit breakers; grounding; lightning protection; battery safety; electrical code compliance

T0B   Antenna safety: tower safety and grounding; erecting an antenna support; safely installing an antenna

T0C   RF hazards: radiation exposure; proximity to antennas; recognized safe power levels; exposure to others; radiation types; duty cycle

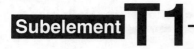

FCC Rules

SUBELEMENT T1
FCC Rules

Your Technician exam (Element 2) will consists of 35 questions taken from the Technician question pool as prepared by the Volunteer Examiner Coordinator's Question Pool Committee. A certain number of questions are taken from each of the 10 subelements. There will be 6 questions from the FCC Rules subelement shown in this chapter. The questions are divided into 6 groups T1A through T1F.

The correct answer (A, B, C or D) is given in bold at the beginning of an explanation section that follows the question and the possible responses. This convention will be used throughout this book. In addition, at the end of each explanation you'll find the page number where this question is discussed in ARRL's *Ham Radio License Manual,* like this: [*Ham Radio License Manual,* page 7-3].

You'll often see a reference to Part 97 of the Federal Communications Commission rules set in brackets, like this: [97.3(a)(4)]. This tells you where to find the exact wording of the Rules as they relate to that question. You'll find the complete Part 97 Rules on the ARRL website at **www.arrl.org.**

SUBELEMENT T1 — FCC Rules, descriptions, and definitions for the Amateur Radio Service, operator and station license responsibilities
[6 exam questions — 6 groups]

T1A — Amateur Radio Service: purpose and permissible use of the Amateur Radio Service, operator/primary station license grant; Meanings of basic terms used in FCC rules; Interference; RACES rules; Phonetics; Frequency Coordinator

T1A01   Which of the following is a purpose of the Amateur Radio Service as stated in the FCC rules and regulations?

A. Providing personal radio communications for as many citizens as possible

B. Providing communications for international non-profit organizations

C. Advancing skills in the technical and communication phases of the radio art

D. All of these choices are correct

[97.1] — Part 97.1 of the FCC's rules is the Basis and Purpose for the Amateur Service:

The rules and regulations in this part are designed to provide an amateur radio service having a fundamental purpose as expressed in the following principles:

(a) Recognition and enhancement of the value of the amateur service to the public as a voluntary noncommercial communication service, particularly with respect to providing emergency communications.

(b) Continuation and extension of the amateur's proven ability to contribute to the advancement of the radio art.

(c) Encouragement and improvement of the amateur service through rules which provide for advancing skills in both the communication and technical phases of the art.

(d) Expansion of the existing reservoir within the amateur radio service of trained operators, technicians, and electronics experts.

(e) Continuation and extension of the amateur's unique ability to enhance international goodwill.

[*Ham Radio License Manual,* page 7-2]

T1A02   Which agency regulates and enforces the rules for the Amateur Radio Service in the United States?

A. FEMA
B. Homeland Security
C. The FCC
D. All of these choices are correct

● [97.1] — The *Federal Communications Commission* or *FCC* is responsible for regulating telecommunications in the United States and all of its possessions. [*Ham Radio License Manual*, page 7-2]

T1A03   What are the FCC rules regarding the use of a phonetic alphabet for station identification in the Amateur Radio Service?

A. It is required when transmitting emergency messages
B. It is prohibited
C. It is required when in contact with foreign stations
D. It is encouraged

● [97.119(b)(2)] — You are required to identify in English, even if you are communicating in a language other than English. The FCC recommends the use of phonetics when you identify by voice — that avoids confusing letters that sound alike. The standard phonetics are words in the English language. You may also identify by CW even if using phone. [*Ham Radio License Manual*, page 8-4]

ITU Phonetic Alphabet

Letter	Word	Pronunciation	Letter	Word	Pronunciation
A	Alfa	**AL** FAH	N	November	NO **VEM** BER
B	Bravo	**BRAH** VOH	O	Oscar	**OSS** CAH
C	Charlie	**CHAR** LEE	P	Papa	PAH **PAH**
D	Delta	**DELL** TAH	Q	Quebec	KEH **BECK**
E	Echo	**ECK** OH	R	Romeo	**ROW** ME OH
F	Foxtrot	**FOKS** TROT	S	Sierra	SEE **AIR** RAH
G	Golf	GOLF	T	Tango	**TANG** GO
H	Hotel	HOH **TELL**	U	Uniform	**YOU** NEE FORM
I	India	**IN** DEE AH	V	Victor	**VIK** TAH
J	Juliet	**JEW** LEE ETT	W	Whiskey	**WISS** KEY
K	Kilo	**KEY** LOH	X	X-Ray	**ECKS** RAY
L	Lima	**LEE** MAH	Y	Yankee	**YANG** KEY
M	Mike	MIKE	Z	Zulu	**ZOO** LOO

T1A04 How many operator/primary station license grants may be
held by any one person?

A. One
B. No more than two
C. One for each band on which the person plans to operate
D. One for each permanent station location from which the person plans
to operate

A [97.5(b)(1)] — An operator license gives you permission to operate an
amateur station according to the rules of the amateur service. The station
license authorizes you to have an amateur station. The combined license is
an *amateur operator/primary station license*. Each person can have only one
such license. [*Ham Radio License Manual*, page 7-3]

T1A05 What is proof of possession of an FCC-issued operator/
primary license grant?

A. A printed operator/primary station license issued by the FCC must be
displayed at the transmitter site
B. The control operator must have an operator/primary station license in
his or her possession when in control of a transmitter
C. The control operator's operator/primary station license must appear
in the FCC ULS consolidated licensee database
D. All of these choices are correct

C [97.7] — Once your information appears in the FCC ULS consolidated
database, that's proof you have been granted an operator/station license and
are fully authorized to go on the air. The FCC no longer routinely issues
printed licenses, although they are available upon request. [*Ham Radio
License Manual*, page 7-5]

T1A06 What is the FCC Part 97 definition of a beacon?

A. A government transmitter marking the amateur radio band edges
B. A bulletin sent by the FCC to announce a national emergency
C. An amateur station transmitting communications for the purposes of
observing propagation or related experimental activities
D. A continuous transmission of weather information authorized in the
amateur bands by the National Weather Service

C [97.3(a)(9)] — Beacon stations are restricted to certain sub-bands to
keep them from causing interference since they transmit under automatic
control and do not listen for other activity. [*Ham Radio License Manual*, page
7-12]

T1A07 **What is the FCC Part 97 definition of a space station?**

A. Any satellite orbiting the earth
B. A manned satellite orbiting the earth
C. An amateur station located more than 50 km above the Earth's surface
D. An amateur station using amateur radio satellites for relay of signals

C [97.3(a)(41)] — The definition is for an amateur station operating in space, not the more common name of the International Space Station (ISS). There is an amateur station on the ISS, so there is a space station operating on the space station! [*Ham Radio License Manual*, page 6-23]

T1A08 **Which of the following entities recommends transmit/receive channels and other parameters for auxiliary and repeater stations?**

A. Frequency Spectrum Manager appointed by the FCC
B. Volunteer Frequency Coordinator recognized by local amateurs
C. FCC Regional Field Office
D. International Telecommunications Union

B [97.3(a)(22)] — A committee of volunteers known as a *frequency coordinator* recommends transmit and receive frequencies. The frequency coordinator representatives are selected by the local or regional amateurs whose stations are eligible to be auxiliary or repeater stations. [*Ham Radio License Manual*, page 7-13]

T1A09 **Who selects a Frequency Coordinator?**

A. The FCC Office of Spectrum Management and Coordination Policy
B. The local chapter of the Office of National Council of Independent Frequency Coordinators
C. Amateur operators in a local or regional area whose stations are eligible to be repeater or auxiliary stations
D. FCC Regional Field Office

C [97.3(a)(22)] — See question T1A08. [*Ham Radio License Manual*, page 7-13]

T1A10 **Which of the following describes the Radio Amateur Civil Emergency Service (RACES)?**

A. A radio service using amateur frequencies for emergency management or civil defense communications
B. A radio service using amateur stations for emergency management or civil defense communications
C. An emergency service using amateur operators certified by a civil defense organization as being enrolled in that organization
D. All of these choices are correct

D [97.3(a)(38), 97.407] — RACES is a special part of the Amateur service created by the FCC to provide communications assistance to local, state, or federal government emergency management agencies during civil emergencies. See Part 97.407 of the FCC rules for more information on RACES. [*Ham Radio License Manual*, page 6-18]

T1A11 **When is willful interference to other amateur radio stations permitted?**

A. To stop another amateur station which is breaking the FCC rules
B. At no time
C. When making short test transmissions
D. At any time, stations in the Amateur Radio Service are not protected from willful interference

B [97.101(d)] — Intentionally creating harmful interference is called *willful interference* and is never allowed. [*Ham Radio License Manual*, page 8-6]

T1B — Authorized frequencies: frequency allocations; ITU; emission modes; restricted sub-bands; spectrum sharing; transmissions near band edges; contacting the International Space Station; power output

T1B01   What is the International Telecommunications Union (ITU)?

    A.   An agency of the United States Department of Telecommunications Management

    B.   A United Nations agency for information and communication technology issues

    C.   An independent frequency coordination agency

    D.   A department of the FCC

B   The *International Telecommunication Union* (ITU) was formed as an agency of the United Nations (UN) in 1949. The ITU is an administrative forum for working out international telecommunications treaties and laws, including frequency allocations. The ITU also maintains international radio laws that all UN countries agree to abide by. [*Ham Radio License Manual,* page 7-14]

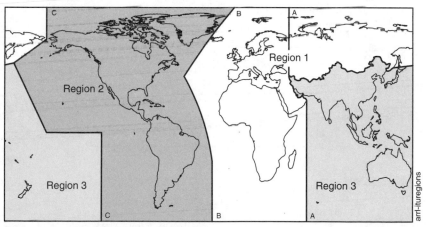

This map shows the world divided into three International Telecommunication Union (ITU) Regions. The US is part of Region 2.

T1B02   Which amateur radio stations may make contact with an amateur radio station on the International Space Station (ISS) using 2 meter and 70 cm band frequencies?

A. Only members of amateur radio clubs at NASA facilities
B. Any amateur holding a Technician or higher-class license
C. Only the astronaut's family members who are hams
D. Contacts with the ISS are not permitted on amateur radio frequencies

B   [97.301, 97.207(c)] — It is only necessary to be licensed to transmit on the uplink frequencies to contact any space station. [*Ham Radio License Manual*, page 6-23]

T1B03   Which frequency is within the 6 meter amateur band?

A. 49.00 MHz
B. 52.525 MHz
C. 28.50 MHz
D. 222.15 MHz

B   [97.301(a)] — The table below shows the Technician VHF/UHF frequency privileges that you are expected to know for your license exam. Remember that you can use the formula f (in MHz) = 300 / wavelength (in meters) or wavelength (in meters) = 300 / f (in MHz) to convert between frequency and wavelength. [*Ham Radio License Manual*, page 7-9]

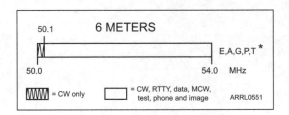

Most Popular VHF and UHF Technician Amateur Bands

Band (Wavelength)   *Frequency Limits*

VHF Range

Band (Wavelength)	Frequency Limits
6 meters	50 – 54 MHz
2 meters	144 – 148 MHz
1.25 meters	219 – 220 MHz
1.25 meters	222 – 225 MHz

UHF Range

Band (Wavelength)	Frequency Limits
70 centimeters	420 – 450 MHz
33 centimeters	902 – 928 MHz
23 centimeters	1240 – 1300 MHz
13 centimeters	2300 – 2310 MHz
13 centimeters	2390 – 2450 MHz

T1B04   **Which amateur band are you using when your station is transmitting on 146.52 MHz?**

A. 2 meter band
B. 20 meter band
C. 14 meter band
D. 6 meter band

A   [97.301(a)] — See question T1B03. [*Ham Radio License Manual,* page 7-9]

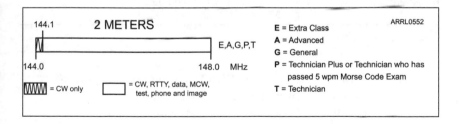

T1B05 **What is the limitation for emissions on the frequencies between 219 and 220 MHz?**

A. Spread spectrum only
B. Fixed digital message forwarding systems only
C. Emergency traffic only
D. Fast-scan television only

B [97.305(c)] — The segment of the 1.25 meter band from 219 to 220 MHz is restricted to digital message forwarding by fixed stations and systems. [*Ham Radio License Manual*, page 7-12]

T1B06 **On which HF bands does a Technician class operator have phone privileges?**

A. None
B. 10 meter band only
C. 80 meter, 40 meter, 15 meter and 10 meter bands
D. 30 meter band only

B [97.301(e), 97.305] [*Ham Radio License Manual*, page 7-11]

Technician HF Privileges (200 watts PEP maximum output)

Band (Wavelength)	Frequency (MHz)
80 meters	3.525-3.600 (CW only)
40 meters	7.025-7.125 (CW only)
15 meters	21.025-21.200 (CW only)
10 meters	28.000-28.300 (CW, RTTY and data)
	28.300-28.500 (CW and SSB)

T1B07 **Which of the following VHF/UHF frequency ranges are limited to CW only?**

A. 50.0 MHz to 50.1 MHz and 144.0 MHz to 144.1 MHz
B. 219 MHz to 220 MHz and 420.0 MHz to 420.1 MHz
C. 902.0 MHz to 902.1 MHZ
D. All of these choices are correct

A [97.305(a),(c)] — There is a small CW-only sub-band occupying the bottom 100 kHz of the 6 and 2 meter bands. [*Ham Radio License Manual*, page 7-12]

T1B08 Which of the following is a result of the fact that the Amateur Radio Service is secondary in all or portions of some amateur bands (such as portions of the 70 cm band)?

A. U.S. amateurs may find non-amateur stations in those portions, and must avoid interfering with them
B. U.S. amateurs must give foreign amateur stations priority in those portions
C. International communications are not permitted in those portions
D. Digital transmissions are not permitted in those portions

A [97.303] — A primary service is *protected* from harmful interference by signals from secondary services. The secondary service gains access to the frequencies in the allocation with the understanding that it must not cause harmful interference to primary service users and it must accept interference from primary users. [*Ham Radio License Manual*, page 7-13]

T1B09 Why should you not set your transmit frequency to be exactly at the edge of an amateur band or sub-band?

A. To allow for calibration error in the transmitter frequency display
B. So that modulation sidebands do not extend beyond the band edge
C. To allow for transmitter frequency drift
D. All of these choices are correct

D [97.101(a), 97.301(a-e)] — Amateurs are allowed to use any frequency in a band, but have to be careful when operating near the edge of the band. All of the signal must be inside the band. Since radios display the *carrier frequency*, remember to leave room for the signal's sidebands. [*Ham Radio License Manual*, page 5-7]

T1B10 Which of the following HF bands have frequencies available to the Technician class operator for RTTY and data transmissions?

A. 10 meter, 12 meter, 17 meter, and 40 meter bands
B. 10 meter, 15 meter, 40 meter, and 80 meter bands
C. 30 meter band only
D. 10 meter band only

D [97.301(e), 97.305(c)] — See the table with question T1B06. [*Ham Radio License Manual*, page 7-11]

T1B11　　What is the maximum peak envelope power output for
Technician class operators using their assigned portions of
the HF bands?

A.　200 watts
B.　100 watts
C.　50 watts
D.　10 watts

A　[97.313] — Below 30 MHz, Novice and Technician licensees are limited
to 200 watts PEP on HF bands. With a few specific restrictions Technicians
are allowed the full legal limit of 1500 watts PEP output above 30 MHz.
[*Ham Radio License Manual*, page 7-12]

T1B12　　Except for some specific restrictions, what is the maximum
peak envelope power output for Technician class operators
using frequencies above 30 MHz?

A.　50 watts
B.　100 watts
C.　500 watts
D.　1500 watts

D　[97.313(b)] — See question T1B11. [*Ham Radio License Manual*, page
7-12]

T1C — Operator licensing: operator classes; sequential and vanity call sign systems; international communications; reciprocal operation; places where the Amateur Radio Service is regulated by the FCC; name and address on FCC license database; license term; renewal; grace period

T1C01 For which license classes are new licenses currently available from the FCC?

A. Novice, Technician, General, Advanced
B. Technician, Technician Plus, General, Advanced
C. Novice, Technician Plus, General, Advanced
D. Technician, General, Amateur Extra

D [97.9(a), 97.17(a)] — There are three other license classes — the Novice, Technician Plus, and Advanced. No new licenses are being granted for these classes. [*Ham Radio License Manual*, page 7-3]

T1C02 Who may select a desired call sign under the vanity call sign rules?

A. Only a licensed amateur with a General or Amateur Extra class license
B. Only a licensed amateur with an Amateur Extra class license
C. Only a licensed amateur who has been licensed continuously for more than 10 years
D. Any licensed amateur

D [97.19] — Licensed hams can pick any available call authorized for their license class. There are many available calls for Technician licensees to choose from in the Group C (1-by-3) and Group D (2-by-3) call sign formats. [*Ham Radio License Manual*, page 7-17]

T1C03 What types of international communications is an FCC-licensed amateur radio station permitted to make?

A. Communications incidental to the purposes of the Amateur Radio Service and remarks of a personal character
B. Communications incidental to conducting business or remarks of a personal nature
C. Only communications incidental to contest exchanges, all other communications are prohibited
D. Any communications that would be permitted by an international broadcast station

A [97.117] — Unless specifically prohibited by the government of either country, any ham can talk to any other ham. International communications must be limited to the purposes of the amateur service or remarks of a personal nature. [*Ham Radio License Manual*, page 7-15]

T1C04 When are you allowed to operate your amateur station in a foreign country?

A. When the foreign country authorizes it
B. When there is a mutual agreement allowing third party communications
C. When authorization permits amateur communications in a foreign language
D. When you are communicating with non-licensed individuals in another country

A [97.107] — To operate at all, the foreign country must permit amateur operation. In addition, you must have permission and when you are inside a country's national boundaries, including territorial waters, you are required to operate according to their rules. You may also operate from any vessel or craft that is documented or registered in the United States. [*Ham Radio License Manual*, page 7-15]

T1C05 Which of the following is a valid call sign for a Technician class amateur radio station?

A. K1XXX
B. KA1X
C. W1XX
D. All of these choices are correct

A See question T1C02. [*Ham Radio License Manual*, page 7-17]

T1C06 From which of the following locations may an FCC-licensed amateur station transmit?

A. From within any country that belongs to the International Telecommunications Union
B. From within any country that is a member of the United Nations
C. From anywhere within International Telecommunications Union (ITU) Regions 2 and 3
D. From any vessel or craft located in international waters and documented or registered in the United States

D [97.5(a)(2)] — See question T1C04. [*Ham Radio License Manual*, page 7-15]

T1C07   What may result when correspondence from the FCC is returned as undeliverable because the grantee failed to provide and maintain a correct mailing address with the FCC?

A. Fine or imprisonment
B. Revocation of the station license or suspension of the operator license
C. Require the licensee to be re-examined
D. A reduction of one rank in operator class

B   [97.23] — The FCC requires you to provide and maintain a valid current mailing address in their database at all times so you can be contacted by mail, if needed. If you move or even change P.O. boxes, be sure to update your information using the FCC ULS online system. [*Ham Radio License Manual*, page 7-8]

T1C08   What is the normal term for an FCC-issued primary station/ operator amateur radio license grant?

A. Five years
B. Life
C. Ten years
D. Twenty years

C   [97.25] — You can renew them indefinitely without ever taking another exam. You can renew online by using the FCC's Universal Licensing System (ULS). Up until 90 days before your license expires, you can also fill out a paper FCC Form 605 and mail it to the FCC. If your license expires, you have a two-year grace period to apply for a new license without taking the exam again. Until your license is renewed, stop transmitting because your license is not valid after it expires. [*Ham Radio License Manual*, page 7-5]

T1C09   What is the grace period following the expiration of an amateur license within which the license may be renewed?

A. Two years
B. Three years
C. Five years
D. Ten years

A   [97.21(a)(b)] — See question T1C08. [*Ham Radio License Manual*, page 7-5]

T1C10 How soon after passing the examination for your first amateur radio license may you operate a transmitter on an Amateur Radio Service frequency?

A. Immediately
B. 30 days after the test date
C. As soon as your operator/station license grant appears in the FCC's license database
D. You must wait until you receive your license in the mail from the FCC

C [97.5a] — See question T1A05. [*Ham Radio License Manual*, page 7-5]

T1C11 If your license has expired and is still within the allowable grace period, may you continue to operate a transmitter on Amateur Radio Service frequencies?

A. No, transmitting is not allowed until the FCC license database shows that the license has been renewed
B. Yes, but only if you identify using the suffix GP
C. Yes, but only during authorized nets
D. Yes, for up to two years

A [97.21(b)] — See question T1C08. [*Ham Radio License Manual,* page 7-5]

T1D — Authorized and prohibited transmission: communications with other countries; music; exchange of information with other services; indecent language; compensation for use of station; retransmission of other amateur signals; codes and ciphers; sale of equipment; unidentified transmissions; one-way transmission

T1D01 With which countries are FCC-licensed amateur radio stations prohibited from exchanging communications?

A. Any country whose administration has notified the International Telecommunications Union (ITU) that it objects to such communications

B. Any country whose administration has notified the American Radio Relay League (ARRL) that it objects to such communications

C. Any country engaged in hostilities with another country

D. Any country in violation of the War Powers Act of 1934

A [97.111(a)(1)] — Some countries do not recognize Amateur Radio, although the number is very small. The FCC can prohibit contacts between US citizens and those of specific other countries by notifying the ITU of its objections. [*Ham Radio License Manual*, page 7-15]

T1D02 Under which of the following circumstances may an amateur radio station make one-way transmissions?

A. Under no circumstances

B. When transmitting code practice, information bulletins, or transmissions necessary to provide emergency communications

C. At any time, as long as no music is transmitted

D. At any time, as long as the material being transmitted did not originate from a commercial broadcast station

B [97.113(b), 97.111(b)] — Broadcasting consists of *one-way transmissions* intended for reception by the general public. Hams are not permitted to make this type of transmission except for the purposes of transmitting code practice, information bulletins for other amateurs, or when necessary for emergency communications. The prohibition on broadcasting includes repeating and relaying transmissions from other communications services. Hams are also specifically prohibited from assisting and participating in news gathering by broadcasting organizations. [*Ham Radio License Manual*, page 8-12]

T1D03   When is it permissible to transmit messages encoded to hide their meaning?

A. Only during contests
B. Only when operating mobile
C. Only when transmitting control commands to space stations or radio control craft
D. Only when frequencies above 1280 MHz are used

C   [97.211(b), 97.215(b), 97.114(a)(4)] — Translating information into data for transmission is called *encoding*. Recovering the encoded information is called *decoding*. Reducing the size of a message to transmit it more efficiently is called *compression*. Most forms of encoding and compression are okay because they are done according to a published protocol. Any ham can look up the protocol and develop the appropriate capabilities to receive and decode data sent with that protocol. Encoding that uses codes or ciphers to hide the meaning of the transmitted message is called *encryption*. Recovering the encrypted information is called *decryption*. Amateurs may not use encryption techniques except for radio control and control transmissions to space stations where interception or unauthorized transmissions could have serious consequences. [*Ham Radio License Manual,* page 8-11]

T1D04   Under what conditions is an amateur station authorized to transmit music using a phone emission?

A. When incidental to an authorized retransmission of manned spacecraft communications
B. When the music produces no spurious emissions
C. When the purpose is to interfere with an illegal transmission
D. When the music is transmitted above 1280 MHz

A   [97.113(a)(4), 97.113(c)] — Music can only be rebroadcast as part of an authorized rebroadcast of space station transmissions — a rather unusual circumstance. This means that you should turn down a vehicle's audio system when you're transmitting from a mobile station. [*Ham Radio License Manual*, page 8-12]

T1D05 When may amateur radio operators use their stations to notify other amateurs of the availability of equipment for sale or trade?

A. When the equipment is normally used in an amateur station and such activity is not conducted on a regular basis
B. When the asking price is $100.00 or less
C. When the asking price is less than its appraised value
D. When the equipment is not the personal property of either the station licensee or the control operator or their close relatives

A [97.113(a)(3)(ii)] — It is okay to advertise equipment for sale as long as it pertains to Amateur Radio and it's not your regular business. You can also order things over the air, as long as you don't do it regularly or as part of your normal income-making activities. No transmissions related to conducting your business or employer's activities are permitted. Your own personal activities don't count as "business" communications, though. One exception is that teachers can be a control operator of a ham station while using ham radio for instruction, but it must be incidental to their job and can't be the majority of their duties. [*Ham Radio License Manual*, page 8-10]

T1D06 What, if any, are the restrictions concerning transmission of language that may be considered indecent or obscene?

A. The FCC maintains a list of words that are not permitted to be used on amateur frequencies
B. Any such language is prohibited
C. The ITU maintains a list of words that are not permitted to be used on amateur frequencies
D. There is no such prohibition

B [97.113(a)(4)] — While you may encounter stations using such language, it's best to avoid controversial topics and expletives. [*Ham Radio License Manual*, page 8-10]

T1D07 What types of amateur stations can automatically retransmit the signals of other amateur stations?

A. Auxiliary, beacon, or Earth stations
B. Repeater, auxiliary, or space stations
C. Beacon, repeater, or space stations
D. Earth, repeater, or space stations

B [97.113(d)] — Retransmitting the signals of another station is generally prohibited, except when you are relaying messages or digital data from another station. Some types of stations (repeaters, auxiliary stations and space stations) are allowed to automatically retransmit signals on different frequencies or channels. [*Ham Radio License Manual*, page 8-12]

T1D08   In which of the following circumstances may the control operator of an amateur station receive compensation for operating that station?

A. When the communication is related to the sale of amateur equipment by the control operator's employer
B. When the communication is incidental to classroom instruction at an educational institution
C. When the communication is made to obtain emergency information for a local broadcast station
D. All of these choices are correct

B   [97.113(a)(3)(iii)] — See question T1D05. [*Ham Radio License Manual*, page 8-11]

T1D09   Under which of the following circumstances are amateur stations authorized to transmit signals related to broadcasting, program production, or news gathering, assuming no other means is available?

A. Only where such communications directly relate to the immediate safety of human life or protection of property
B. Only when broadcasting communications to or from the space shuttle
C. Only where noncommercial programming is gathered and supplied exclusively to the National Public Radio network
D. Only when using amateur repeaters linked to the internet

A   [97.113(5)(b)] — See question T1D02. [*Ham Radio License Manual*, page 8-12]

T1D10   What is the meaning of the term broadcasting in the FCC rules for the Amateur Radio Service?

A. Two-way transmissions by amateur stations
B. Transmission of music
C. Transmission of messages directed only to amateur operators
D. Transmissions intended for reception by the general public

D   [97.3(a)(10)] — See question T1D02. [*Ham Radio License Manual*, page 8-12]

T1D11 **When may an amateur station transmit without on-the-air identification?**

A. When the transmissions are of a brief nature to make station adjustments
B. When the transmissions are unmodulated
C. When the transmitted power level is below 1 watt
D. When transmitting signals to control model craft

D [97.119(a)] — Unidentified transmissions are not allowed. Unidentified means that no call sign was associated with a transmission. The only exceptions are for space stations, or for when your signals are controlling a model craft. [*Ham Radio License Manual,* page 8-3]

T1E — Control operator and control types: control operator required; eligibility; designation of control operator; privileges and duties; control point; local, automatic and remote control; location of control operator

T1E01 **When is an amateur station permitted to transmit without a control operator?**

A. When using automatic control, such as in the case of a repeater
B. When the station licensee is away and another licensed amateur is using the station
C. When the transmitting station is an auxiliary station
D. Never

D [97.7(a)] — A *control operator* is the licensed amateur designated to be responsible for making sure that transmissions comply with FCC rules. That doesn't have to be the same person as the station owner. Any licensed amateur can be a control operator.

The *station licensee* is responsible for designating the control operator. A control operator must be named in the FCC amateur license database or have reciprocal operating authorization. All transmissions must be made under the supervision of a control operator and there can be only one control operator for a station at a time. [*Ham Radio License Manual*, page 8-1]

T1E02 **Who may be the control operator of a station communicating through an amateur satellite or space station?**

A. Only an Amateur Extra Class operator
B. A General class or higher licensee who has a satellite operator certification
C. Only an Amateur Extra Class operator who is also an AMSAT member
D. Any amateur whose license privileges allow them to transmit on the satellite uplink frequency

D [97.301, 97.207(c)] — Satellite contacts, including contacts with the amateur station on the International Space Station, can be made by any amateur licensed to transmit on the *uplink* frequency. [*Ham Radio License Manual*, page 6-22]

T1E03 **Who must designate the station control operator?**

A. The station licensee
B. The FCC
C. The frequency coordinator
D. The ITU

A [97.103(b)] — See question T1E01. [*Ham Radio License Manual*, page 8-1]

T1E04 **What determines the transmitting privileges of an amateur station?**

A. The frequency authorized by the frequency coordinator
B. The frequencies printed on the license grant
C. The highest class of operator license held by anyone on the premises
D. The class of operator license held by the control operator

D [97.103(b)] — As the control operator, you may operate the station in any way permitted by the privileges of your license class. It doesn't matter what the station owner's privileges are, only the privileges of the control operator. When acting as a control operator, you are restricted to the privileges of your license class. [*Ham Radio License Manual*, page 8-2]

T1E05 **What is an amateur station control point?**

A. The location of the station's transmitting antenna
B. The location of the station transmitting apparatus
C. The location at which the control operator function is performed
D. The mailing address of the station licensee

C [97.3(a)(14)] — See question T1E01. [*Ham Radio License Manual*, page 8-1]

T1E06   When, under normal circumstances, may a Technician class licensee be the control operator of a station operating in an exclusive Amateur Extra class operator segment of the amateur bands?

A. At no time
B. When operating a special event station
C. As part of a multi-operator contest team
D. When using a club station whose trustee is an Amateur Extra class operator licensee

A   [97.301] — See question T1E04. [*Ham Radio License Manual*, page 8-2]

T1E07   When the control operator is not the station licensee, who is responsible for the proper operation of the station?

A. All licensed amateurs who are present at the operation
B. Only the station licensee
C. Only the control operator
D. The control operator and the station licensee are equally responsible

D   [97.103(a)] — The station owner is responsible for limiting access to the station only to responsible licensees who will follow the FCC rules. Note that the FCC will presume the station licensee to be the control operator unless there is a written record to the contrary. [*Ham Radio License Manual*, page 8-2]

T1E08   Which of the following is an example of automatic control?

A. Repeater operation
B. Controlling the station over the internet
C. Using a computer or other device to send CW automatically
D. Using a computer or other device to identify automatically

A   [97.3(a)(6), 97.205(d)] — There are three different ways that a operator can control a transmitter:

Local control — a control operator is physically present at the control point. This is the situation for nearly all amateur stations, including mobile operation. Any type of station can be locally controlled.

Remote control — the control point is located away from the transmitter and the control operator adjusts or operates the transmitter indirectly via some kind of control link. The control operator must be present at the control point during all transmission. Many stations operate under remote control over an Internet link. Any station can be remotely controlled.

Automatic control — *t*he station operates completely under the control of devices and procedures that ensure compliance with FCC rules. A control operator is still required, but need not be at the control point when the station is transmitting. Repeaters, beacons and space stations are allowed to be automatically controlled. Digipeaters that relay messages, such as for the APRS network, are also automatically controlled.

[*Ham Radio License Manual,* page 8-9]

T1E09 Which of the following is true of remote control operation?

A. The control operator must be at the control point
B. A control operator is required at all times
C. The control operator indirectly manipulates the controls
D. All of these choices are correct

D [97.109(c)] — See question T1E08. [*Ham Radio License Manual*, page 8-9]

T1E10 Which of the following is an example of remote control as defined in Part 97?

A. Repeater operation
B. Operating the station over the internet
C. Controlling a model aircraft, boat, or car by amateur radio
D. All of these choices are correct

B [97.3(a)(39)] — See question T1E08. [*Ham Radio License Manual*, page 8-9]

T1E11 Who does the the FCC presume to be the control operator of an amateur station, unless documentation to the contrary is in the station records?

A. The station custodian
B. The third-party participant
C. The person operating the station equipment
D. The station licensee

D [97.103(a)] — See question T1E07. [*Ham Radio License Manual,* page 8-2]

T1F — Station identification; repeaters; third-party communications; club stations; FCC inspection

T1F01 When must the station licensee make the station and its records available for FCC inspection?

A. At any time ten days after notification by the FCC of such an inspection
B. At any time upon request by an FCC representative
C. Only after failing to comply with an FCC notice of violation
D. Only when presented with a valid warrant by an FCC official or government agent

B [97.103(c)] — As a federal licensee, you are obligated to make your station available for inspection upon request by an FCC representative. By accepting the FCC rules and regulations for the amateur service, you agree that your station could be inspected any time. [*Ham Radio License Manual*, page 7-8]

T1F02   When using tactical identifiers such as Race Headquarters during a community service net operation, how often must your station transmit the station's FCC-assigned call sign?

A. Never, the tactical call is sufficient
B. Once during every hour
C. At the end of each communication and every ten minutes during a communication
D. At the end of every transmission

C   [97.119(a)] — Tactical call signs (or tactical IDs) are used to help identify where a station is and what it is doing. Tactical calls don't replace regular call signs and the regular identification rules apply — give your FCC-assigned call sign every 10 minutes and at the end of the communication. [*Ham Radio License Manual,* page 8-4]

T1F03   When is an amateur station required to transmit its assigned call sign?

A. At the beginning of each contact, and every 10 minutes thereafter
B. At least once during each transmission
C. At least every 15 minutes during and at the end of a communication
D. At least every 10 minutes during and at the end of a communication

D   [97.119(a)] — Give your call sign at least once every 10 minutes during a contact and when the communication is finished. The communication can include contacts with several stations, such as during a net or contest. [*Ham Radio License Manual,* page 8-3]

T1F04   Which of the following is an acceptable language to use for station identification when operating in a phone sub-band?

A. Any language recognized by the United Nations
B. Any language recognized by the ITU
C. The English language
D. English, French, or Spanish

C   [97.119(b)(2)] — See question T1A03. [*Ham Radio License Manual,* page 8-4]

T1F05   What method of call sign identification is required for a station transmitting phone signals?

A. Send the call sign followed by the indicator RPT
B. Send the call sign using a CW or phone emission
C. Send the call sign followed by the indicator R
D. Send the call sign using only a phone emission

B   [97.119(b)(2)] — See question T1A03. [*Ham Radio License Manual,* page 8-4]

T1F06   Which of the following formats of a self-assigned indicator is acceptable when identifying using a phone transmission?

A. KL7CC stroke W3
B. KL7CC slant W3
C. KL7CC slash W3
D. All of these choices are correct

D   [97.119(c)] — FCC Part 97.119(c) says, "One or more indicators may be included with the call sign. Each indicator must be separated from the call sign by the slant mark (/) or by any suitable word that denotes the slant mark. If an indicator is self-assigned, it must be included before, after, or both before and after, the call sign." [*Ham Radio License Manual, page 8-5*]

T1F07   Which of the following restrictions apply when a non-licensed person is allowed to speak to a foreign station using a station under the control of a Technician class control operator?

A. The person must be a U.S. citizen
B. The foreign station must be one with which the U.S. has a third-party agreement
C. The licensed control operator must do the station identification
D. All of these choices are correct

B   [97.115(a)(2)] — The exact definition of third-party communication is a message from an amateur station control operator to another amateur station control operator on behalf of another person. That "other person" is the "third party." Simplifying the definition, any time that you send or receive information via ham radio on behalf of any unlicensed person or an organization, even if the person is right there in the station with you — that's third-party communications. [*Ham Radio License Manual*, page 8-8]

The following table shows which countries have *third-party agreements* with the United States. If the other country isn't on this list, third-party communication with that country is not permitted.

Third-Party Agreements

The United States has third-party agreements with the following nations.

V2	Antigua/Barbuda
LO-LW	Argentina
VK	Australia
V3	Belize
CP	Bolivia
E7	Bosnia-Herzegovina
PP-PY	Brazil
VE, VO, VY	Canada
CA-CE	Chile
HJ-HK	Colombia
D6	Comoros (Federal Islamic Republic of)
TI, TE	Costa Rica
CM, CO	Cuba
HI	Dominican Republic
J7	Dominica
HC-HD	Ecuador
YS	El Salvador
C5	Gambia, The
9G	Ghana
J3	Grenada
TG	Guatemala
8R	Guyana
HH	Haiti
HQ-HR	Honduras
4X, 4Z	Israel
6Y	Jamaica
JY	Jordan
EL	Liberia
V7	Marshall Islands
XA-XI	Mexico
V6	Micronesia, Federated States of
YN	Nicaragua
HO-HP	Panama
ZP	Paraguay
OA-OC	Peru
DU-DZ	Philippines
VR6	Pitcairn Island
V4	St. Kitts/Nevis
J6	St. Lucia
J8	St. Vincent and the Grenadines
9L	Sierra Leone
ZR-ZU	South Africa
3DA	Swaziland
9Y-9Z	Trinidad/Tobago
TA-TC	Turkey
GB	United Kingdom
CV-CX	Uruguay
YV-YY	Venezuela
4U1ITU	ITU — Geneva
4U1VIC	VIC — Vienna

T1F08   **What is meant by the term Third Party Communications?**

A. A message from a control operator to another amateur station control operator on behalf of another person
B. Amateur radio communications where three stations are in communications with one another
C. Operation when the transmitting equipment is licensed to a person other than the control operator
D. Temporary authorization for an unlicensed person to transmit on the amateur bands for technical experiments

A   [97.3(a)(47)] — See question T1F07. [*Ham Radio License Manual*, page 8-7]

T1F09   **What type of amateur station simultaneously retransmits the signal of another amateur station on a different channel or channels?**

A. Beacon station
B. Earth station
C. Repeater station
D. Message forwarding station

C   [97.3(a)(40)] — Repeaters consist of a receiver and transmitter that re-transmit the information from a received signal simultaneously on another frequency or channel [*Ham Radio License Manual*, page 2-8]

T1F10   **Who is accountable should a repeater inadvertently retransmit communications that violate the FCC rules?**

A. The control operator of the originating station
B. The control operator of the repeater
C. The owner of the repeater
D. Both the originating station and the repeater owner

A   [97.205(g)] — Repeater owners must ensure the repeater operates in compliance with FCC rules. Repeater users are responsible for proper operation via the repeater, however. [*Ham Radio License Manual*, page 8-9]

T1F11   **Which of the following is a requirement for the issuance of a club station license grant?**

A. The trustee must have an Amateur Extra class operator license grant
B. The club must have at least four members
C. The club must be registered with the American Radio Relay League
D. All of these choices are correct

B   [97.5(b)(2)] — Each club must have a licensed *trustee* who actually holds the club license and is designated by a club officer. Clubs must have at least four members and be organized as in rule 97.5(b). [*Ham Radio License Manual*, page 7-3]

Operating Procedures

SUBELEMENT T2 — Operating Procedures
[3 Exam Questions — 3 Groups]

T2A — Station operation: choosing an operating frequency; calling another station; test transmissions; procedural signs; use of minimum power; choosing an operating frequency; band plans; calling frequencies; repeater offsets

T2A01 Which of the following is a common repeater frequency offset in the 2 meter band?
A. Plus or minus 5 MHz
B. Plus or minus 600 kHz
C. Plus or minus 500 kHz
D. Plus or minus 1 MHz

B [*Ham Radio License Manual,* page 6-12]

Standard Repeater Offsets by Band

Band	Offset
10 meters	−100 kHz
6 meters	Varies by region: −500 kHz, −1 MHz, −1.7 MHz
2 meters	+ or −600 kHz
1.25 meters	−1.6 MHz
70 cm	+ or −5 MHz
902 MHz	12 MHz
1296 MHz	12 MHz

T2A02 What is the national calling frequency for FM simplex operations in the 2 meter band?
A. 146.520 MHz
B. 145.000 MHz
C. 432.100 MHz
D. 446.000 MHz

A 146.52 MHz is the standard 2 meter simplex calling frequency. 446.00 MHz is the simplex calling frequency on the 70 cm band. [*Ham Radio License Manual,* page 6-6]

T2A03   What is a common repeater frequency offset in the 70 cm band?

    A. Plus or minus 5 MHz
    B. Plus or minus 600 kHz
    C. Plus or minus 500 kHz
    D. Plus or minus 1 MHz

A   See the table with question T2A01. [*Ham Radio License Manual*, page 6-12]

T2A04   What is an appropriate way to call another station on a repeater if you know the other station's call sign?

    A. Say break, break, then say the station's call sign
    B. Say the station's call sign, then identify with your call sign
    C. Say CQ three times, then the other station's call sign
    D. Wait for the station to call CQ, then answer it

B   If you want to respond to a station asking for a call or want to contact a station whose call sign you already know, just say the other station's call sign once, followed by "this is" or "from," then give your call sign. [*Ham Radio License Manual*, page 6-4]

T2A05   How should you respond to a station calling CQ?

    A. Transmit CQ followed by the other station's call sign
    B. Transmit your call sign followed by the other station's call sign
    C. Transmit the other station's call sign followed by your call sign
    D. Transmit a signal report followed by your call sign

C   If you hear a station calling CQ, send the CQing station's call sign once then yours once or twice. Send your call clearly and distinctly so that it can be understood if there is noise or interference. [*Ham Radio License Manual*, page 6-7]

T2A06   Which of the following is required when making on-the-air test transmissions?

    A. Identify the transmitting station
    B. Conduct tests only between 10 p.m. and 6 a.m. local time
    C. Notify the FCC of the transmissions
    D. All of these choices are correct

A   Identification rules apply to on-the-air test transmissions, as well, no matter how brief. The call sign must be given once every 10 minutes and at the end of transmissions. [*Ham Radio License Manual*, page 8-5]

T2A07   What is meant by repeater offset?

A. The difference between a repeater's transmit frequency and its receive frequency
B. The repeater has a time delay to prevent interference
C. The repeater station identification is done on a separate frequency
D. The number of simultaneous transmit frequencies used by a repeater

A   The difference between repeater input and output frequencies is called the repeater's *offset* or *shift*. See also questions T2A01 and T2A03. [*Ham Radio License Manual*, page 6-12]

T2A08   What is the meaning of the procedural signal CQ?

A. Call on the quarter hour
B. A new antenna is being tested (no station should answer)
C. Only the called station should transmit
D. Calling any station

D   CQ is a procedural signal that means "I am calling any station." [*Ham Radio License Manual*, page 6-6]

T2A10   What brief statement indicates that you are listening on a repeater and looking for a contact?

A. The words Hello test followed by your call sign
B. Your call sign
C. The repeater call sign followed by your call sign
D. The letters QSY followed by your call sign

B   Because repeaters usually have a strong signal on a known frequency, it's not necessary to make a long transmission to attract listeners tuning by. The easiest way is to just give your call sign. [*Ham Radio License Manual*, page 6-4]

T2A10   What is a band plan, beyond the privileges established by the FCC?

A. A voluntary guideline for using different modes or activities within an amateur band
B. A mandated list of operating schedules
C. A list of scheduled net frequencies
D. A plan devised by a club to indicate frequency band usage

A   Amateurs created band plans that help organize the different modes and activities. It's important to keep in mind that band plans are voluntary agreements designed for normal conditions. They are not regulations and amateurs are expected to be flexible. [*Ham Radio License Manual*, page 6-2]

T2A11 **What term describes an amateur station that is transmitting and receiving on the same frequency?**

A. Full duplex
B. Diplex
C. Simplex
D. Multiplex

C *Simplex* and *duplex* refer to the type of communication taking place. Simplex communication is transmitting and receiving on the same frequency. Duplex communication uses one frequency for transmitting and another frequency for receiving. [*Ham Radio License Manual*, page 6-2]

T2A12 **Which of the following is a guideline when choosing an operating frequency for calling CQ?**

A. Listen first to be sure that no one else is using the frequency
B. Ask if the frequency is in use
C. Make sure you are in your assigned band
D. All of these choices are correct

D Before you call CQ you should do three things:

• Be sure the frequency is one your license privileges authorize you to use!

• Listen to be sure the frequency is not already in use. If you don't hear any signals in five to ten seconds the frequency may be available.

• Make a short transmission asking if the frequency is in use.

[*Ham Radio License Manual*, page 6-7]

T2B — VHF/UHF operating practices: SSB phone; FM repeater; simplex; splits and shifts; CTCSS; DTMF; tone squelch; carrier squelch; phonetics; operational problem resolution; Q signals

T2B01 **What is the most common use of the reverse split function of a VHF/UHF transceiver?**

A. Reduce power output
B. Increase power output
C. Listen on a repeater's input frequency
D. Listen on a repeater's output frequency

C Many radios have a reverse split function that swaps the transmit and receive frequencies. This enables you to listen for the other station on the repeater's input frequency. [*Ham Radio License Manual*, page 6-6]

T2B02   **What term describes the use of a sub-audible tone transmitted along with normal voice audio to open the squelch of a receiver?**

A. Carrier squelch
B. Tone burst
C. DTMF
D. CTCSS

D   Repeater access tones are known by various names: Continuous Tone Coded Squelch System (CTCSS), PL (for Private Line, the Motorola trade name) or sub-audible. [*Ham Radio License Manual*, page 6-13]

T2B03   **If a station is not strong enough to keep a repeater's receiver squelch open, which of the following might allow you to receive the station's signal?**

A. Open the squelch on your radio
B. Listen on the repeater input frequency
C. Listen on the repeater output frequency
D. Increase your transmit power

B   Listening on the repeater's input frequency is often helpful when a weak station is trying to access the repeater, but isn't quite strong enough to hold the repeater's squelch open. [*Ham Radio License Manual*, page 6-6]

T2B04   **Which of the following could be the reason you are unable to access a repeater whose output you can hear?**

A. Improper transceiver offset
B. The repeater may require a proper CTCSS tone from your transceiver
C. The repeater may require a proper DCS tone from your transceiver
D. All of these choices are correct

D   If you can hear a repeater's signal and you're sure you are using the right offset, but can't access the repeater, then you probably don't have your radio set up to use the right type or frequency of access tone. [*Ham Radio License Manual*, page 6-13]

T2B05   **What might be the problem if a repeater user says your transmissions are breaking up on voice peaks?**

A. You have the incorrect offset
B. You need to talk louder
C. You are talking too loudly
D. Your transmit power is too high

C   An overmodulated FM signal has excessive deviation and is said to be *overdeviating*. Overdeviation is usually caused by speaking too loudly into the microphone and may cause interference on adjacent channels. It often generates noise or distortion on voice peaks, called "breaking up." To reduce overdeviation, speak more softly or move the microphone farther from your mouth. [*Ham Radio License Manual*, page 5-8]

T2B06 **What type of tones are used to control repeaters linked by the Internet Relay Linking Project (IRLP) protocol?**
A. DTMF
B. CTCSS
C. EchoLink
D. Sub-audible

A To initiate a contact or exercise a control function on an IRLP-linked repeater, a control code is used. The code is a sequence of DTMF (Dual-tone Multi-Frequency) tones, like dialing a phone number. [*Ham Radio License Manual*, page 6-15]

T2B07 **How can you join a digital repeater's talk group?**
A. Register your radio with the local FCC office
B. Join the repeater owner's club
C. Program your radio with the group's ID or code
D. Sign your call after the courtesy tone

C The central DMR controller organizes users of the DMR network into talk groups. Each talk group has an ID or code. By programming your radio with those IDs and codes, you can join the group and your audio will be shared with all other members of the group. Talk groups allow groups of users to share a channel at different times without being heard by other users on the channel. [*Ham Radio License Manual*, page 6-15]

T2B08 **Which of the following applies when two stations transmitting on the same frequency interfere with each other?**
A. Common courtesy should prevail, but no one has absolute right to an amateur frequency
B. Whoever has the strongest signal has priority on the frequency
C. Whoever has been on the frequency the longest has priority on the frequency
D. The station that has the weakest signal has priority on the frequency

A If a transmission seriously degrades, obstructs or repeatedly interrupts the communications of a regulated service, that's considered harmful interference. Common courtesy should prevail but remember that no one has an absolute right to any frequency. [*Ham Radio License Manual*, page 8-6]

T2B09 **What is a talk group on a DMR digital repeater?**
A. A group of operators sharing common interests
B. A way for groups of users to share a channel at different times without being heard by other users on the channel
C. A protocol that increases the signal-to-noise ratio when multiple repeaters are linked together
D. A net that meets at a particular time

B See question T2B07. [*Ham Radio License Manual*, page 6-15]

T2B10 Which Q signal indicates that you are receiving interference from other stations?

A. QRM
B. QRN
C. QTH
D. QSB

A [*Ham Radio License Manual,* page 6-7]

Q-Signals

These Q-signals are the ones used most often on the air. (Q abbreviations take the form of questions only when they are sent followed by a question mark.)

QRG	Your exact frequency (or that of _____) is _____kHz.
	Will you tell me my exact frequency (or that of _____)?
QRL	I am busy (or I am busy with _____). Are you busy? Usually used to see if a frequency is busy.
QRM	Your transmission is being interfered with _____
	(1. Nil; 2. Slightly; 3. Moderately; 4. Severely; 5. Extremely.)
	Is my transmission being interfered with?
QRN	I am troubled by static _____. (1 to 5 as under QRM.)
	Are you troubled by static?
QRO	Increase power. Shall I increase power?
QRP	Decrease power. Shall I decrease power?
QRQ	Send faster (_____wpm). Shall I send faster?
QRS	Send more slowly (_____wpm). Shall I send more slowly?
QRT	Stop sending. Shall I stop sending?
QRU	I have nothing for you. Have you anything for me?
QRV	I am ready. Are you ready?
QRX	I will call you again at _____hours (on _____kHz).
	When will you call me again? Minutes are usually implied rather than hours.
QRZ	You are being called by _____ (on _____kHz).
	Who is calling me?
QSB	Your signals are fading. Are my signals fading?
QSK	I can hear you between signals; break in on my transmission.
	Can you hear me between your signals and if so can I break in on your transmission?
QSL	I am acknowledging receipt.
	Can you acknowledge receipt (of a message or transmission)?
QSO	I can communicate with _____ direct (or relay through _____).
	Can you communicate with _____ direct or by relay?
QSP	I will relay to _____. Will you relay to _____?
QST	General call preceding a message addressed to all amateurs and ARRL members. This is in effect "CQ ARRL."
QSX	I am listening to _____ on _____kHz. Will you listen to _____on _____kHz?
QSY	Change to transmission on another frequency (or on _____kHz).
	Shall I change to transmission on another frequency (or on _____kHz)?
QTC	I have _____messages for you (or for _____).
	How many messages have you to send?
QTH	My location is _____. What is your location?
QTR	The time is _____. What is the correct time?

T2B11 Which Q signal indicates that you are changing frequency?

A. QRU
B. QSY
C. QSL
D. QRZ

B See question T2B10. [*Ham Radio License Manual*, page 6-7]

T2B12 Why are simplex channels designated in the VHF/UHF band plans?

A. So that stations within mutual communications range can communicate without tying up a repeater
B. For contest operation
C. For working DX only
D. So that stations with simple transmitters can access the repeater without automated offset

A Since only so many repeaters can share a band in a region, hams must use them wisely. If two stations are in range of each other, they can try simplex communication to avoid occupying or "tying up" a repeater. [*Ham Radio License Manual*, page 6-6]

T2B13 Where may SSB phone be used in amateur bands above 50 MHz?

A. Only in sub-bands allocated to General class or higher licensees
B. Only on repeaters
C. In at least some portion of all these bands
D. On any band as long as power is limited to 25 watts

C Every amateur band at and above 50 MHz has frequencies available for CW and SSB operation. [*Ham Radio License Manual*, page 6-2]

T2B14 Which of the following describes a linked repeater network?

A. A network of repeaters where signals received by one repeater are repeated by all the repeaters
B. A repeater with more than one receiver
C. Multiple repeaters with the same owner
D. A system of repeaters linked by APRS

A Linked repeaters share the signals each receives and retransmit them. Linking can be performed via a digital connection such as over the internet or by direct radio links between the repeaters. [*Ham Radio License Manual*, page 6-12]

T2C — Public service: emergency and non-emergency operations; applicability of FCC rules; RACES and ARES; net and traffic procedures; operating restrictions during emergencies

T2C01   When do the FCC rules NOT apply to the operation of an amateur station?

A. When operating a RACES station
B. When operating under special FEMA rules
C. When operating under special ARES rules
D. Never, FCC rules always apply

D   [97.103(a)] — You are bound by FCC rules at all times, even if using your radio in support of a public safety agency. [*Ham Radio License Manual*, page 6-18]

T2C02   What is meant by the term NCS used in net operation?

A. Nominal Control System
B. Net Control Station
C. National Communications Standard
D. Normal Communications Syntax

B   The *Net Control Station* (NCS) is in charge of net operation and directs the operation of other stations participating in the net. [*Ham Radio License Manual*, page 6-16]

T2C03   What should be done when using voice modes to ensure that voice messages containing unusual words are received correctly?

A. Send the words by voice and Morse code
B. Speak very loudly into the microphone
C. Spell the words using a standard phonetic alphabet
D. All of these choices are correct

C   To help a receiving station copy a message exactly, proper names (such as "John Doe") and unusual words (such as material names or model identifiers) are spelled out using standard phonetics. [*Ham Radio License Manual*, page 6-17]

T2C04   What do RACES and ARES have in common?

   A. They represent the two largest ham clubs in the United States
   B. Both organizations broadcast road and weather information
   C. Neither may handle emergency traffic supporting public service agencies
   D. Both organizations may provide communications during emergencies

D   The two largest Amateur Radio emergency response organizations are ARES® (Amateur Radio Emergency Service®) sponsored by the ARRL and the Radio Amateur Civil Emergency Service (RACES). Both organizations provide emergency and disaster response communications. [*Ham Radio License Manual*, page 6-18]

T2C05   What does the term traffic refer to in net operation?

   A. Formal messages exchanged by net stations
   B. The number of stations checking in and out of a net
   C. Operation by mobile or portable stations
   D. Requests to activate the net by a served agency

A   Messages passed over the air that follow a set, formal structure are called *traffic*. Exchanging messages is called *traffic handling*. [*Ham Radio License Manual*, page 6-15]

T2C06   Which of the following is an accepted practice to get the immediate attention of a net control station when reporting an emergency?

   A. Repeat SOS three times followed by the call sign of the reporting station
   B. Press the push-to-talk button three times
   C. Begin your transmission by saying Priority or Emergency followed by your call sign
   D. Play a pre-recorded emergency alert tone followed by your call sign

C   A station with emergency traffic should break into an ongoing contact or net at any time. If you are operating on phone and are reporting an emergency, break in by saying "Priority" or "Emergency," followed by your call sign. [*Ham Radio License Manual*, page 6-16]

T2C07   **Which of the following is an accepted practice for an amateur operator who has checked into a net?**

A. Provided that the frequency is quiet, announce the station call sign and location every 5 minutes
B. Move 5 kHz away from the net's frequency and use high power to ask other hams to keep clear of the net frequency
C. Remain on frequency without transmitting until asked to do so by the net control station
D. All of these choices are correct

C   Once you have checked into a formal net under the direction of an NCS, it is important to not disrupt the net. Do not transmit unless you are specifically requested or authorized to do so or a request is made for capabilities or information that you can provide. [*Ham Radio License Manual*, page 6-16]

T2C08   **Which of the following is a characteristic of good traffic handling?**

A. Passing messages exactly as received
B. Making decisions as to whether messages are worthy of relay or delivery
C. Ensuring that any newsworthy messages are relayed to the news media
D. All of these choices are correct

A   The most important job for traffic handlers during emergency and disaster net operation is the ability to accurately relay or "pass" messages exactly as written, spoken or received. [*Ham Radio License Manual*, page 6-16]

T2C09   **Are amateur station control operators ever permitted to operate outside the frequency privileges of their license class?**

A. No
B. Yes, but only when part of a FEMA emergency plan
C. Yes, but only when part of a RACES emergency plan
D. Yes, but only if necessary in situations involving the immediate safety of human life or protection of property

D   In an emergency situation where there is immediate risk to life or property and normal forms of communication are unavailable, you may use any means possible to address that risk, including operating outside the frequency privileges of your license. [*Ham Radio License Manual*, page 6-18]

T2C10   **What information is contained in the preamble of a formal traffic message?**

   A. The email address of the originating station
   B. The address of the intended recipient
   C. The telephone number of the addressee
   D. The information needed to track the message

D   The preamble is made up of several bits of information about the message. These establish a unique identity for each message so that it can be handled and tracked appropriately as it moves through the Amateur Radio traffic handling system.

• Number — a unique number assigned by the station that creates the radiogram

• Precedence — a description of the nature of the radiogram: Routine, Priority, Emergency and Welfare

• Handling Instructions (HX) — for special instructions in how to the handle the radiogram.

• Station of Origin — the call sign of the radio station from which the radiogram was first sent by Amateur Radio. (This allows information about the message to be returned to the sending station.)

• Check — the number of words and word equivalents in the radiogram text.

• Place of Origin — the name of the town from which the radiogram started

• Time and Date — the time and date the radiogram is received at the station that first sent it

• Address — the complete name, street and number, city and state to whom the radiogram is going

Following the preamble is the text of the radiogram.

[*Ham Radio License Manual,* page 6-17]

T2C11   **What is meant by the term check, in reference to a formal traffic message?**

   A. The number of words or word equivalents in the text portion of the message
   B. The value of a money order attached to the message
   C. A list of stations that have relayed the message
   D. A box on the message form that indicates that the message was received and/or relayed

A   See question T2C10. [*Ham Radio License Manual,* page 6-17]

T2C12   What is the Amateur Radio Emergency Service (ARES)?

A. Licensed amateurs who have voluntarily registered their qualifications and equipment for communications duty in the public service
B. Licensed amateurs who are members of the military and who voluntarily agreed to provide message handling services in the case of an emergency
C. A training program that provides licensing courses for those interested in obtaining an amateur license to use during emergencies
D. A training program that certifies amateur operators for membership in the Radio Amateur Civil Emergency Service

A ARES consists of licensed amateurs who have voluntarily registered their qualifications and equipment for communications duty in the public service. [*Ham Radio License Manual*, page 6-18]

Radio Wave Characteristics

SUBELEMENT T3 — Radio wave characteristics: properties of radio waves; propagation modes
[3 Exam Questions — 3 Groups]

T3A — Radio wave characteristics: how a radio signal travels; fading; multipath; polarization; wavelength vs absorption; antenna orientation

T3A01 What should you do if another operator reports that your station's 2 meter signals were strong just a moment ago, but now they are weak or distorted?

A. Change the batteries in your radio to a different type
B. Turn on the CTCSS tone
C. Ask the other operator to adjust his squelch control
D. Try moving a few feet or changing the direction of your antenna if possible, as reflections may be causing multi-path distortion

D Moving your antenna just a few feet may avoid the location at which cancellation is occurring due to multipath propagation. [*Ham Radio License Manual*, page 4-2]

T3A02 Why might the range of VHF and UHF signals be greater in the winter?

A. Less ionospheric absorption
B. Less absorption by vegetation
C. Less solar activity
D. Less tropospheric absorption

B Vegetation can also absorb VHF and UHF radio waves. This can result in greater range in the winter. [*Ham Radio License Manual*, page 4-2]

T3A03   **What antenna polarization is normally used for long-distance weak-signal CW and SSB contacts using the VHF and UHF bands?**

A. Right-hand circular
B. Left-hand circular
C. Horizontal
D. Vertical

C   Horizontal polarization is preferred for weak-signal contacts because it results in lower ground losses when the wave reflects from or travels along the ground. [*Ham Radio License Manual*, page 4-16]

T3A04   **What can happen if the antennas at opposite ends of a VHF or UHF line of sight radio link are not using the same polarization?**

A. The modulation sidebands might become inverted
B. Signals could be significantly weaker
C. Signals have an echo effect on voices
D. Nothing significant will happen

B   When the polarizations of transmit and receive antennas aren't aligned the same, the received signal can be dramatically reduced. [*Ham Radio License Manual*, page 4-6]

T3A05   **When using a directional antenna, how might your station be able to access a distant repeater if buildings or obstructions are blocking the direct line of sight path?**

A. Change from vertical to horizontal polarization
B. Try to find a path that reflects signals to the repeater
C. Try the long path
D. Increase the antenna SWR

B   On VHF and UHF, if a direct signal path is blocked by building or other obstruction, a beam antenna can be used to aim the signal at a reflecting surface to bypass the obstruction. [*Ham Radio License Manual*, page 4-16]

T3A06   **What term is commonly used to describe the rapid fluttering sound sometimes heard from mobile stations that are moving while transmitting?**

A. Flip-flopping
B. Picket fencing
C. Frequency shifting
D. Pulsing

B   Because "dead spots" from multipath are usually spaced about ½-wavelength apart, VHF or UHF signals from a station in motion can take on a rapid variation in strength known as *mobile flutter* or *picket-fencing*. [*Ham Radio License Manual*, page 4-3]

T3A07 What type of wave carries radio signals between transmitting and receiving stations?

A. Electromagnetic
B. Electrostatic
C. Surface acoustic
D. Ferromagnetic

A Also see question T3B03. [*Ham Radio License Manual*, page 4-6]

T3A08 Which of the following is a likely cause of irregular fading of signals received by ionospheric reflection?

A. Frequency shift due to Faraday rotation
B. Interference from thunderstorms
C. Random combining of signals arriving via different paths
D. Intermodulation distortion

C Multipath propagation of signals from distant stations results in irregular fading, even when reception is generally good, because signals taking different paths can interfere with each other at the receiving antenna. [*Ham Radio License Manual*, page 4-2]

T3A09 Which of the following results from the fact that skip signals refracted from the ionosphere are elliptically polarized?

A. Digital modes are unusable
B. Either vertically or horizontally polarized antennas may be used for transmission or reception
C. FM voice is unusable
D. Both the transmitting and receiving antennas must be of the same polarization

B Because as a radio wave travels through the ionosphere its polarization changes from vertical or horizontal to a combination of the two, both vertical and horizontal antennas are effective for receiving and transmitting on the HF bands where skip propagation is common. [*Ham Radio License Manual*, page 4-7]

T3A10 What may occur if data signals arrive via multiple paths?

A. Transmission rates can be increased by a factor equal to the number of separate paths observed
B. Transmission rates must be decreased by a factor equal to the number of separate paths observed
C. No significant changes will occur if the signals are transmitted using FM
D. Error rates are likely to increase

D Distortion caused by multipath can cause VHF and UHF digital data signals to be received with a higher error rate, even though the signal may be strong. [*Ham Radio License Manual*, page 4-3]

T3A11   **Which part of the atmosphere enables the propagation of radio signals around the world?**

A. The stratosphere
B. The troposphere
C. The ionosphere
D. The magnetosphere

C   Reflections from the ionosphere allow radio waves to be received hundreds or thousands of miles away. [*Ham Radio License Manual*, page 4-3]

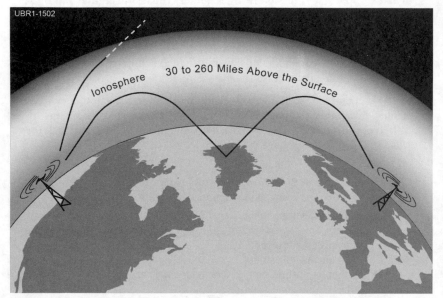

UBR1-1502

Ionosphere   30 to 260 Miles Above the Surface

The ionosphere is formed by solar ultraviolet (UV) radiation. The UV rays knock electrons loose from air molecules, creating weakly charged layers at different heights. These layers can absorb or refract radio signals, sometimes bending them back to the Earth.

T3A12   **How might fog and light rain affect radio range on the 10 meter and 6 meter bands?**

A. Fog and rain absorb these wavelength bands
B. Fog and light rain will have little effect on these bands
C. Fog and rain will deflect these signals
D. Fog and rain will increase radio range

B   Precipitation such as fog and rain can absorb microwave and UHF radio waves although it has little effect at HF and on the lower VHF bands. [*Ham Radio License Manual*, page 4-2]

T3A13 What weather condition would decrease range at microwave frequencies?

A. High winds
B. Low barometric pressure
C. Precipitation
D. Colder temperatures

C See question T3A12. [*Ham Radio License Manual,* page 4-2]

T3B — Radio and electromagnetic wave properties: the electromagnetic spectrum; wavelength vs frequency; nature and velocity of electromagnetic waves; definition of UHF, VHF, HF bands; calculating wavelength

T3B01 What is the name for the distance a radio wave travels during one complete cycle?

A. Wave speed
B. Waveform
C. Wavelength
D. Wave spread

C The *wavelength* of a radio wave is the distance that it travels during one complete cycle. Wavelength is represented by the Greek letter lambda, λ. [*Ham Radio License Manual,* page 2-5]

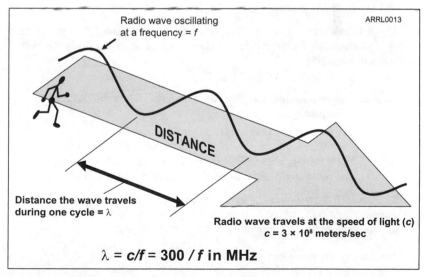

Radio wave oscillating at a frequency = *f*

ARRL0013

DISTANCE

Distance the wave travels during one cycle = λ

Radio wave travels at the speed of light (*c*)
c = 3 × 10⁸ meters/sec

$$\lambda = c/f = 300 \, / \, f \text{ in MHz}$$

As a radio wave travels, it oscillates at the frequency of the signal. The distance covered by the wave during the time it takes for one complete cycle is its wavelength.

T3B02   What property of a radio wave is used to describe its polarization?

A. The orientation of the electric field
B. The orientation of the magnetic field
C. The ratio of the energy in the magnetic field to the energy in the electric field
D. The ratio of the velocity to the wavelength

A   *Polarization* refers to the orientation of the radio wave's electric field. [*Ham Radio License Manual*, page 4-6]

T3B03   What are the two components of a radio wave?

A. AC and DC
B. Voltage and current
C. Electric and magnetic fields
D. Ionizing and non-ionizing radiation

C   An *electromagnetic wave* is a combination of varying electric and magnetic fields. [*Ham Radio License Manual*, page 4-6]

T3B04   How fast does a radio wave travel through free space?

A. At the speed of light
B. At the speed of sound
C. Its speed is inversely proportional to its wavelength
D. Its speed increases as the frequency increases

A   All radio waves travel at the speed of light (represented by a lower-case *c*) in whatever medium they are traveling, such as air. [*Ham Radio License Manual*, page 2-5]

T3B05   How does the wavelength of a radio wave relate to its frequency?

A. The wavelength gets longer as the frequency increases
B. The wavelength gets shorter as the frequency increases
C. There is no relationship between wavelength and frequency
D. The wavelength depends on the bandwidth of the signal

B   Because radio waves travel at a constant speed, $\lambda = c / f$ (λ is one wavelength; c is the speed of light; and f is frequency). This means that as frequency increases, wavelength decreases and vice-versa.

$$\lambda \text{ in meters} = \frac{300}{f \text{ in MHz}}$$

Because of this relationship, amateur bands are often referred to by wavelength.

[*Ham Radio License Manual*, page 2-5]

T3B06 What is the formula for converting frequency to approximate wavelength in meters?

A. Wavelength in meters equals frequency in hertz multiplied by 300
B. Wavelength in meters equals frequency in hertz divided by 300
C. Wavelength in meters equals frequency in megahertz divided by 300
D. Wavelength in meters equals 300 divided by frequency in megahertz

D See question T3B05. [*Ham Radio License Manual*, page 2-6]

T3B07 What property of radio waves is often used to identify the different frequency bands?

A. The approximate wavelength
B. The magnetic intensity of waves
C. The time it takes for waves to travel one mile
D. The voltage standing wave ratio of waves

A See question T3B05. [*Ham Radio License Manual*, page 2-6]

T3B08 What are the frequency limits of the VHF spectrum?

A. 30 to 300 kHz
B. 30 to 300 MHz
C. 300 to 3000 kHz
D. 300 to 3000 MHz

B [*Ham Radio License Manual,* page 2-4]

RF Spectrum Ranges
[T3B08 – T3B10]

Range Name	Abbreviation	Frequency Range
Very Low Frequency	VLF	3 kHz – 30 kHz
Low Frequency	LF	30 kHz – 300 kHz
Medium Frequency	MF	300 kHz – 3 MHz
High Frequency	HF	3 MHz – 30 MHz
Very High Frequency	VHF	30 MHz – 300 MHz
Ultra High Frequency	UHF	300 MHz – 3 GHz
Super High Frequency	SHF	3 GHz – 30 GHz
Extremely High Frequency	EHF	30 GHz – 300 GHz

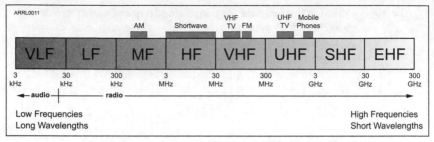

The radio spectrum extends over a very wide range of frequencies. The drawing shows the frequency ranges used by broadcast stations and mobile phones. Amateurs can use small frequency bands in the MF and higher frequency regions of the spectrum.

T3B09 What are the frequency limits of the UHF spectrum?

A. 30 to 300 kHz
B. 30 to 300 MHz
C. 300 to 3000 kHz
D. 300 to 3000 MHz

D See the table for question T3B08. [*Ham Radio License Manual*, page 2-4]

T3B10 What frequency range is referred to as HF?

A. 300 to 3000 MHz
B. 30 to 300 MHz
C. 3 to 30 MHz
D. 300 to 3000 kHz

C See the table for question T3B08. [*Ham Radio License Manual*, page 2-4]

T3B11 What is the approximate velocity of a radio wave as it travels through free space?

A. 150,000 kilometers per second
B. 300,000,000 meters per second
C. 300,000,000 miles per hour
D. 150,000 miles per hour

B The speed of light in space and air is very close to 300 million meters per second (300,000,000 or 3×10^8 meters per second). [*Ham Radio License Manual*, page 2-5]

T3C — Propagation modes: line of sight; sporadic E; meteor and auroral scatter and reflections; tropospheric ducting; F layer skip; radio horizon

T3C01   Why are direct (not via a repeater) UHF signals rarely heard from stations outside your local coverage area?

  A. They are too weak to go very far
  B. FCC regulations prohibit them from going more than 50 miles
  C. UHF signals are usually not reflected by the ionosphere
  D. UHF signals are absorbed by the ionospheric D layer

C   VHF and UHF signals usually pass through the ionosphere and are lost to space. Long-distance ionospheric propagation is the most common way for hams to make long-distance contacts on the HF bands. [*Ham Radio License Manual,* page 4-4]

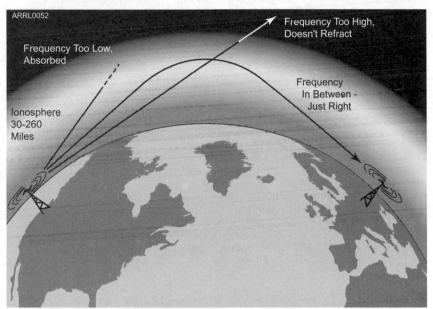

Signals that are too low in frequency are absorbed by the ionosphere and lost. Signals that are too high in frequency pass through the ionosphere and are also lost. Signals in the right range of frequencies are refracted back toward the Earth and are received hundreds or thousands of miles away.

T3C02 **Which of the following is an advantage of HF vs VHF and higher frequencies?**

A. HF antennas are generally smaller
B. HF accommodates wider bandwidth signals
C. Long distance ionospheric propagation is far more common on HF
D. There is less atmospheric interference (static) on HF

C See question T3C01. [*Ham Radio License Manual*, page 4-4]

T3C03 **What is a characteristic of VHF signals received via auroral reflection?**

A. Signals from distances of 10,000 or more miles are common
B. The signals exhibit rapid fluctuations of strength and often sound distorted
C. These types of signals occur only during winter nighttime hours
D. These types of signals are generally strongest when your antenna is aimed west

B Because the aurora is constantly changing, the reflected signals change strength quickly and are often distorted. [*Ham Radio License Manual*, page 4-4]

T3C04 **Which of the following propagation types is most commonly associated with occasional strong over-the-horizon signals on the 10, 6, and 2 meter bands?**

A. Backscatter
B. Sporadic E
C. D layer absorption
D. Gray-line propagation

B Patches of the ionosphere's E layer can become sufficiently ionized to reflect VHF and UHF signals back to Earth. This is called *sporadic E*, *Es*, or *E-skip* propagation. [*Ham Radio License Manual*, page 4-4]

T3C05 **Which of the following effects might cause radio signals to be heard despite obstructions between the transmitting and receiving stations?**

A. Knife-edge diffraction
B. Faraday rotation
C. Quantum tunneling
D. Doppler shift

A Radio waves can be *diffracted* as they travel past sharp edges of large objects. This type of propagation is called *knife-edge diffraction*. [*Ham Radio License Manual*, page 4-2]

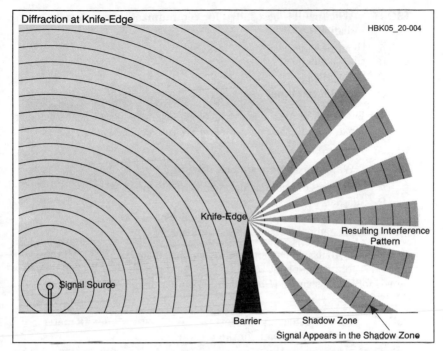

Diffraction at Knife-Edge

HBK05_20-004

Knife-Edge

Resulting Interference Pattern

Signal Source

Barrier

Shadow Zone

Signal Appears in the Shadow Zone

VHF and UHF radio waves are diffracted around the sharp edge of a solid object, such as a building, hill or other obstruction. Some signals appear behind the obstruction as a result of interference between waves at the edge and those farther away. The resulting interference pattern creates shadowed areas where little signal is present.

T3C06 **What mode is responsible for allowing over-the-horizon VHF and UHF communications to ranges of approximately 300 miles on a regular basis?**

A. Tropospheric ducting
B. D layer refraction
C. F2 layer refraction
D. Faraday rotation

A Tropospheric ducting is regularly used by amateurs to make VHF and UHF contacts that would otherwise be impossible by line-of-sight propagation. [*Ham Radio License Manual*, page 4-3]

T3C07   What band is best suited for communicating via meteor scatter?

A. 10 meter band
B. 6 meter band
C. 2 meter band
D. 70 centimeter band

B   The best band for meteor scatter is 6 meters, and contacts can be made at distances up to 1200 to 1500 miles. [*Ham Radio License Manual*, page 4-4]

T3C08   What causes tropospheric ducting?

A. Discharges of lightning during electrical storms
B. Sunspots and solar flares
C. Updrafts from hurricanes and tornadoes
D. Temperature inversions in the atmosphere

D   Weather fronts and temperature inversions create layers of air next to each other form atmospheric *ducts* that can guide VHF, UHF, and microwave signals for long distances. [*Ham Radio License Manual*, page 4-3]

T3C09   What is generally the best time for long-distance 10 meter band propagation via the F layer?

A. From dawn to shortly after sunset during periods of high sunspot activity
B. From shortly after sunset to dawn during periods of high sunspot activity
C. From dawn to shortly after sunset during periods of low sunspot activity
D. From shortly after sunset to dawn during periods of low sunspot activity

A   During the years of maximum solar activity, the upper HF bands, such as 10 meters, are likely to be open from dawn until shortly after sunset. Occasionally, the F layers can even reflect 6 meter (50 MHz) signals at the sunspot cycle's peak. [*Ham Radio License Manual*, page 4-4]

T3C10   Which of the following bands may provide long distance communications during the peak of the sunspot cycle?

A. 6 or 10 meter bands
B. 23 centimeter band
C. 70 centimeter or 1.25 meter bands
D. All of these choices are correct

A   See question T3C09. [*Ham Radio License Manual*, page 4-4]

T3C11   Why do VHF and UHF radio signals usually travel somewhat farther than the visual line of sight distance between two stations?

A. Radio signals move somewhat faster than the speed of light
B. Radio waves are not blocked by dust particles
C. The Earth seems less curved to radio waves than to light
D. Radio waves are blocked by dust particles

C   By bending signals slightly back toward the ground, refraction counteracts the curvature of the Earth and allows signals at these frequencies to be received at distances somewhat beyond the visual horizon. [*Ham Radio License Manual*, page 4-2]

Amateur Radio Practices and Station Set-Up

SUBELEMENT T4 — Amateur radio practices and station set-up
[2 Exam Questions — 2 Groups]

T4A — Station setup: connecting microphones; reducing unwanted emissions; power source; connecting a computer; RF grounding; connecting digital equipment; connecting an SWR meter

T4A01 What must be considered to determine the minimum current capacity needed for a transceiver power supply?

A. Efficiency of the transmitter at full power output
B. Receiver and control circuit power
C. Power supply regulation and heat dissipation
D. All of these choices are correct

D The current rating of a supply must be at least as much as the sum of the maximum current used by everything hooked up to the supply. To determine the maximum current rating needed for a transceiver, you must consider the transmitter's efficiency at full power output, how much current the receiver and control circuits require, and the power supply's regulation and ability to dissipate heat at full load. [*Ham Radio License Manual*, page 5-16]

T4A02 How might a computer be used as part of an amateur radio station?

A. For logging contacts and contact information
B. For sending and/or receiving CW
C. For generating and decoding digital signals
D. All of these choices are correct

D Just about any ham radio function or capability has the potential to involve a computer: bookkeeping chores such as logging contacts, operating on the digital modes, and sending and receiving CW are all often performed by a computer. [*Ham Radio License Manual*, page 5-11]

T4A03 Why should wiring between the power source and radio be heavy-gauge wire and kept as short as possible?

A. To avoid voltage falling below that needed for proper operation
B. To provide a good counterpoise for the antenna
C. To avoid RF interference
D. All of these choices are correct

A If the wire is too thin, its resistance (R) will create a *voltage drop*, $V = I \times R$. Longer wire lengths also increase resistance. The resulting lower voltage at the radio can cause it to operate improperly, possibly distorting your output signal or creating interference. [*Ham Radio License Manual*, page 5-17]

T4A04 Which computer sound card port is connected to a transceiver's headphone or speaker output for operating digital modes?

A. Headphone output
B. Mute
C. Microphone or line input
D. PCI or SDI

C The figure shows an example of how a station is configured to use digital modes. If a standalone TNC is used, it is connected to one of the computer's digital data ports via a COM port or USB interface. The TNC is then connected to the radio's microphone input (for transmit audio) and speaker or headphone output (for receive audio). If a sound card is used instead, its output is connected to the radio's microphone input and the speaker or headphone output is connected to the sound card input. If you use a sound card, you may need a digital communications interface to supply the PTT (push-to-talk) signal for keying the transmitter. [*Ham Radio License Manual*, page 5-15]

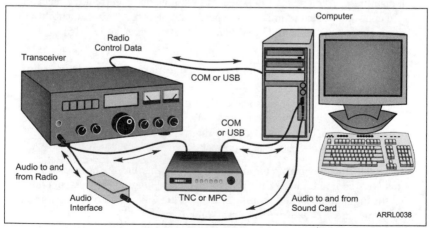

Data interfaces are connected between the transceiver's audio inputs and outputs and the computer's data connections (USB or COM ports) or sound card jacks. A TNC or MPC (multi-protocol controller) converts between data and audio. An audio interface isolates the computer sound card from the radio to prevent hum.

T4A05   **What is the proper location for an external SWR meter?**

A. In series with the feed line, between the transmitter and antenna
B. In series with the station's ground
C. In parallel with the push-to-talk line and the antenna
D. In series with the power supply cable, as close as possible to the radio

A   An SWR meter is placed in series with the feed line, usually right at the output of the radio. [*Ham Radio License Manual*, page 4-18]

T4A06   **Which of the following connections might be used between a voice transceiver and a computer for digital operation?**

A. Receive and transmit mode, status, and location
B. Antenna and RF power
C. Receive audio, transmit audio, and push-to-talk (PTT)
D. NMEA GPS location and DC power

C   See question T4A04. [*Ham Radio License Manual*, page 5-15]

T4A07   **How is a computer's sound card used when conducting digital communications?**

A. The sound card communicates between the computer CPU and the video display
B. The sound card records the audio frequency for video display
C. The sound card provides audio to the radio's microphone input and converts received audio to digital form
D. All of these choices are correct

C   See question T4A04. [*Ham Radio License Manual*, page 5-15]

T4A08   **Which of the following conductors provides the lowest impedance to RF signals?**

A. Round stranded wire
B. Round copper-clad steel wire
C. Twisted-pair cable
D. Flat strap

D   Use short, wide conductors such as copper flashing or strap or heavy solid wire (#8 AWG or larger). Solid strap is best because it presents the lowest impedance to RF. [*Ham Radio License Manual*, page 9-7]

T4A09 Which of the following could you use to cure distorted audio caused by RF current on the shield of a microphone cable?

A. Band-pass filter
B. Low-pass filter
C. Preamplifier
D. Ferrite choke

D *RF choke* or *common-mode* filters made of *ferrite* material are used to reduce RF currents flowing on unshielded wires such as speaker cables, ac power cords, and telephone modular cords. *Ferrite chokes* are also used to reduce RF current on the outside of shielded audio, microphone, and computer cables. [*Ham Radio License Manual*, page 9-8]

T4A10 What is the source of a high-pitched whine that varies with engine speed in a mobile transceiver's receive audio?

A. The ignition system
B. The alternator
C. The electric fuel pump
D. Anti-lock braking system controllers

B Vehicle power wiring often carries a significant amount of noise that can affect your radio's operation. *Alternator whine* is caused by noise on the dc power system inside your own vehicle. You might hear it with the received audio but more likely it will be heard by others as a high-pitched whine on your transmitted audio that varies with your engine speed. [*Ham Radio License Manual*, page 5-17]

T4A11 Where should the negative return connection of a mobile transceiver's power cable be connected?

A. At the battery or engine block ground strap
B. At the antenna mount
C. To any metal part of the vehicle
D. Through the transceiver's mounting bracket

A Connect the radio's negative lead to the negative battery terminal or where the battery ground lead is connected to the vehicle body. [*Ham Radio License Manual*, page 5-17]

T4B — Operating controls: tuning; use of filters; squelch function; AGC; memory channels; transceiver operation

T4B01 What may happen if a transmitter is operated with the microphone gain set too high?

A. The output power might be too high
B. The output signal might become distorted
C. The frequency might vary
D. The SWR might increase

B Excessive modulation results in distortion of transmitted speech and unwanted or *spurious* transmitter outputs on adjacent frequencies where they cause interference. *Overmodulation* of an AM or SSB signal is caused by speaking too loudly or by setting the microphone gain or speech compression too high, possibly resulting in distortion of the transmitted signal. [*Ham Radio License Manual*, page 5-8]

T4B02 Which of the following can be used to enter the operating frequency on a modern transceiver?

A. The keypad or VFO knob
B. The CTCSS or DTMF encoder
C. The Automatic Frequency Control
D. All of these choices are correct

A The control used for tuning is the *VFO*, an abbreviation for *variable frequency oscillator*. In older radios, the VFO knob changes the frequency of an oscillator circuit that in turn determines the radio's operating frequency. In most current radios, the VFO control is read by a microprocessor that controls the radio's frequency. Some radios also have a keypad that you can use to enter frequencies directly. [*Ham Radio License Manual*, page 5-6]

T4B03 What is the purpose of the squelch control on a transceiver?

A. To set the highest level of volume desired
B. To set the transmitter power level
C. To adjust the automatic gain control
D. To mute receiver output noise when no signal is being received

D To keep from having to listen to continuous noise when no signal is present, the *squelch* circuit mutes the receiver's audio output when no signal is present. [*Ham Radio License Manual*, page 5-9]

T4B04 **What is a way to enable quick access to a favorite frequency on your transceiver?**

A. Enable the CTCSS tones
B. Store the frequency in a memory channel
C. Disable the CTCSS tones
D. Use the scan mode to select the desired frequency

B *Memories* or *memory channels* are used to store frequencies and modes for later recall. Memories are provided so that you can quickly tune to frequently used or favorite frequencies. [*Ham Radio License Manual*, page 5-7]

T4B05 **Which of the following would reduce ignition interference to a receiver?**

A. Change frequency slightly
B. Decrease the squelch setting
C. Turn on the noise blanker
D. Use the RIT control

C Special circuits are employed to get rid of noise or at least limit its effect. A *noise blanker* (NB) senses the sharp pulses from arcing power lines, motors, or vehicle ignition systems and temporarily mutes the receiver during the pulse. [*Ham Radio License Manual*, page 5-10]

T4B06 **Which of the following controls could be used if the voice pitch of a single-sideband signal seems too high or low?**

A. The AGC or limiter
B. The bandwidth selection
C. The tone squelch
D. The receiver RIT or clarifier

D See question T4B07. [*Ham Radio License Manual*, page 5-10]

T4B07 **What does the term RIT mean?**

A. Receiver Input Tone
B. Receiver Incremental Tuning
C. Rectifier Inverter Test
D. Remote Input Transmitter

B *Receiver incremental tuning* (RIT) is a fine-tuning control used for SSB or CW operation. RIT allows the operator to adjust the receiver frequency without changing the transmitter frequency. This allows you to tune in a station that is slightly off frequency or to adjust the pitch of an operator's voice that seems too high or low. [*Ham Radio License Manual*, page 5-10]

T4B08 What is the advantage of having multiple receive bandwidth choices on a multimode transceiver?

A. Permits monitoring several modes at once
B. Permits noise or interference reduction by selecting a bandwidth matching the mode
C. Increases the number of frequencies that can be stored in memory
D. Increases the amount of offset between receive and transmit frequencies

B Having multiple filters available allows you to reduce noise or interference by selecting a filter with just enough bandwidth to pass the desired signal. [*Ham Radio License Manual*, page 5-9]

T4B09 Which of the following is an appropriate receive filter bandwidth for minimizing noise and interference for SSB reception?

A. 500 Hz
B. 1000 Hz
C. 2400 Hz
D. 5000 Hz

C A receiver rejects unwanted signals through the use of *filters*. A "narrow" filter has a smaller bandwidth and a "wide" filter has greater bandwidth. Wide filters (around 2.4 kHz bandwidth) are used for SSB reception and narrow filters (around 500 Hz) are used for CW reception. [*Ham Radio License Manual*, page 5-9]

T4B10 Which of the following is an appropriate receive filter bandwidth for minimizing noise and interference for CW reception?

A. 500 Hz
B. 1000 Hz
C. 2400 Hz
D. 5000 Hz

A See question T4B09. [*Ham Radio License Manual*, page 5-9]

T4B11 What is the function of automatic gain control, or AGC?

A. To keep received audio relatively constant
B. To protect an antenna from lightning
C. To eliminate RF on the station cabling
D. An asymmetric goniometer control used for antenna matching

A A receiver's *automatic gain control (AGC)* circuitry constantly adjusts the receiver's sensitivity to keep the output volume constant for both weak and strong signals. [*Ham Radio License Manual*, page 5-9]

T4B12 Which of the following could be used to remove power line noise or ignition noise?

A. Squelch
B. Noise blanker
C. Notch filter
D. All of these choices are correct

B See question T4B05. [*Ham Radio License Manual*, page 5-10]

T4B13 Which of the following is a use for the scanning function of an FM transceiver?

A. To check incoming signal deviation
B. To prevent interference to nearby repeaters
C. To scan through a range of frequencies to check for activity
D. To check for messages left on a digital bulletin board

C The *scanning* function of your radio can be used to listen for activity on a range of frequencies, such as repeater or simplex channels. [*Ham Radio License Manual*, page 6-10]

Electrical Principles

SUBELEMENT T5 — Electrical principles: math for electronics; electronic principles; Ohm's Law
[4 exam questions — 4 groups]

T5A — Electrical principles, units, and terms: current and voltage; conductors and insulators; alternating and direct current; series and parallel circuits

T5A01 Electrical current is measured in which of the following units?

A. Volts
B. Watts
C. Ohms
D. Amperes

D Current is measured in units of *amperes* which is abbreviated as A or amps. [*Ham Radio License Manual*, page 3-1]

T5A02 Electrical power is measured in which of the following units?

A. Volts
B. Watts
C. Ohms
D. Amperes

B Power is measured in *watts* which are abbreviated as W. [*Ham Radio License Manual*, page 3-7]

T5A03 What is the name for the flow of electrons in an electric circuit?

A. Voltage
B. Resistance
C. Capacitance
D. Current

D Electric *current* (represented in equations by the symbol I or i) is the flow of *electrons*. [*Ham Radio License Manual*, page 3-1]

T5A04 **What is the name for a current that flows only in one direction?**

A. Alternating current
B. Direct current
C. Normal current
D. Smooth current

B Current that flows in one direction all the time is *direct current,* abbreviated *dc*. [*Ham Radio License Manual*, page 3-2]

T5A05 **What is the electrical term for the electromotive force (EMF) that causes electron flow?**

A. Voltage
B. Ampere-hours
C. Capacitance
D. Inductance

A *Voltage* (represented in equations by the symbol E or e) is the *electromotive force* or *electric potential* that makes electrons move. [*Ham Radio License Manual*, page 3-1]

T5A06 **How much voltage does a mobile transceiver typically require?**

A. About 12 volts
B. About 30 volts
C. About 120 volts
D. About 240 volts

A Radios that operate from a "12 V" supply usually work best at the slightly higher voltage of 13.8 V typical of vehicle power systems with the engine running. [*Ham Radio License Manual*, page 5-16]

T5A07 **Which of the following is a good electrical conductor?**

A. Glass
B. Wood
C. Copper
D. Rubber

C Materials such as copper in which electrons flow easily in response to an applied voltage are *conductors*. [*Ham Radio License Manual*, page 3-5]

T5A08   Which of the following is a good electrical insulator?

A. Copper
B. Glass
C. Aluminum
D. Mercury

B   Materials such as glass and ceramics, dry wood and paper, most plastics, and other non-metals that resist or prevent the flow of electrons are *insulators*. [*Ham Radio License Manual*, page 3-5]

T5A09   What is the name for a current that reverses direction on a regular basis?

A. Alternating current
B. Direct current
C. Circular current
D. Vertical current

A   Current that regularly reverses direction is *alternating current*, abbreviated *ac*. [*Ham Radio License Manual*, page 3-2]

T5A10   Which term describes the rate at which electrical energy is used?

A. Resistance
B. Current
C. Power
D. Voltage

C   *Power*, represented by the symbol *P*, is the rate at which electrical energy is used. [*Ham Radio License Manual*, page 3-7]

T5A11   What is the unit of electromotive force?

A. The volt
B. The watt
C. The ampere
D. The ohm

A   Voltage is measured in units of *volts* which are abbreviated as V. [*Ham Radio License Manual*, page 3-2]

T5A12   **What describes the number of times per second that an alternating current makes a complete cycle?**

A. Pulse rate
B. Speed
C. Wavelength
D. Frequency

D   Like a water wave, a radio wave continually varies in strength or *amplitude* like the *sine wave* shown in the figure. This continual change is called *oscillating*. As the signal oscillates, each complete up-and-down sequence is called a *cycle*. The number of *cycles per second* is the signal's *frequency*, represented by a lower-case *f*. The unit of frequency is hertz (Hz). [*Ham Radio License Manual,* page 2-3]

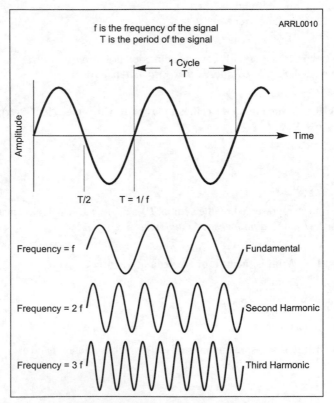

The frequency of a signal and its period are reciprocals. Higher frequency means shorter period and vice-versa.

T5A13 **In which type of circuit is current the same through all components?**

A. Series
B. Parallel
C. Resonant
D. Branch

A   If two or more components such as light bulbs are connected in a circuit so that the same current must flow through all of them, that is a *series* circuit. [*Ham Radio License Manual*, page 3-2]

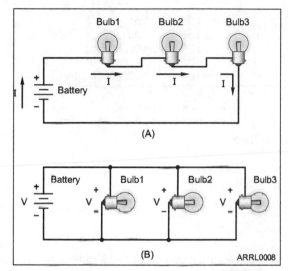

Part A shows three light bulbs and a battery connected in a series circuit. The same current flows from the battery through all three light bulbs. Part B shows the same bulbs and battery connected in a parallel circuit. The same voltage from the battery is applied across each light bulb.

T5A14 **In which type of circuit is voltage the same across all components?**

A. Series
B. Parallel
C. Resonant
D. Branch

B   If two or more components are connected so that the same voltage is present across all of them, that is a *parallel* circuit. [*Ham Radio License Manual*, page 3-2]

T5B — Math for electronics: conversion of electrical units; decibels; the metric system

International System of Units (SI) — Metric Units

Prefix	Symbol	Multiplication Factor		
Tera	T	10^{12}	=	1,000,000,000,000
Giga	G	10^{9}	=	1,000,000,000
Mega	M	10^{6}	=	1,000,000
Kilo	k	10^{3}	=	1000
Hecto	h	10^{2}	=	100
Deca	da	10^{1}	=	10
Deci	d	10^{-1}	=	0.1
Centi	c	10^{-2}	=	0.01
Milli	m	10^{-3}	=	0.001
Micro	μ	10^{-6}	=	0.000001
Nano	n	10^{-9}	=	0.000000001
Pico	p	10^{-12}	=	0.000000000001

T5B01 How many milliamperes is 1.5 amperes?

A. 15 milliamperes
B. 150 milliamperes
C. 1500 milliamperes
D. 15,000 milliamperes

C 1.5 A = 1.5 × 1,000 mA per A = 1,500 mA [*Ham Radio License Manual*, page 2-2]

T5B02 What is another way to specify a radio signal frequency of 1,500,000 hertz?

A. 1500 kHz
B. 1500 MHz
C. 15 GHz
D. 150 kHz

A 1,500,000 Hz = 1,500,000 × .000001 MHz per Hz = 1.5 MHz [*Ham Radio License Manual*, page 2-2]

T5B03 How many volts are equal to one kilovolt?

A. One one-thousandth of a volt
B. One hundred volts
C. One thousand volts
D. One million volts

C 1 kV = 1,000 × 1 V = 1,000 V [*Ham Radio License Manual*, page 2-2]

T5B04   How many volts are equal to one microvolt?

A. One one-millionth of a volt
B. One million volts
C. One thousand kilovolts
D. One one-thousandth of a volt

A   1 μV = 0.000001 V = one one-millionth of a volt [*Ham Radio License Manual*, page 2-2]

T5B05   Which of the following is equal to 500 milliwatts?

A. 0.02 watts
B. 0.5 watts
C. 5 watts
D. 50 watts

B   500 mW = 500 × 0.001 W = 0.5 W [*Ham Radio License Manual*, page 2-2]

T5B06   If an ammeter calibrated in amperes is used to measure a 3000-milliampere current, what reading would it show?

A. 0.003 amperes
B. 0.3 amperes
C. 3 amperes
D. 3,000,000 amperes

C   3,000 mA = 3 × 1,000 mA per A = 3 A [*Ham Radio License Manual*, page 2-2]

T5B07   If a frequency display calibrated in megahertz shows a reading of 3.525 MHz, what would it show if it were calibrated in kilohertz?

A. 0.003525 kHz
B. 35.25 kHz
C. 3525 kHz
D. 3,525,000 kHz

C   3.525 MHz = 3.525 × 1,000 kHz per MHz = 3,525 kHz [*Ham Radio License Manual*, page 2-2]

T5B08   How many microfarads are equal to 1,000,000 picofarads?

A. 0.001 microfarads
B. 1 microfarad
C. 1000 microfarads
D. 1,000,000,000 microfarads

B   1,000,000 pF = 1,000,000 × 0.000001 μF per pF = 1 μF [*Ham Radio License Manual*, page 2-2]

T5B09   What is the approximate amount of change, measured in decibels (dB), of a power increase from 5 watts to 10 watts?

A. 2 dB
B. 3 dB
C. 5 dB
D. 10 dB

B $10 \log (10 / 5) = 10 \log (2) = 3$ dB [*Ham Radio License Manual*, page 4-8]

T5B10   What is the approximate amount of change, measured in decibels (dB), of a power decrease from 12 watts to 3 watts?

A. -1 dB
B. -3 dB
C. -6 dB
D. -9 dB

C $10 \log (3 /12) = 10 \log (0.25) = -6$ dB [*Ham Radio License Manual*, page 4-8]

T5B11   What is the amount of change, measured in decibels (dB), of a power increase from 20 watts to 200 watts?

A. 10 dB
B. 12 dB
C. 18 dB
D. 28 dB

A $10 \log (200 / 20) = 10 \log (10) = 10$ dB [*Ham Radio License Manual*, page 4-8]

T5B12   Which of the following frequencies is equal to 28,400 kHz?

A. 28.400 MHz
B. 2.800 MHz
C. 284.00 MHz
D. 28.400 kHz

A   28,400 kHz = 28,400 × 0.001 MHz per kHz = 28.4 MHz [*Ham Radio License Manual*, page 2-2]

T5B13   If a frequency display shows a reading of 2425 MHz, what frequency is that in GHz?

A. 0.002425 GHz
B. 24.25 GHz
C. 2.425 GHz
D. 2425 GHz

C   2,425 MHz = 2,425 × 0.001 GHz per MHz = 2.425 GHz. [*Ham Radio License Manual,* page 2-2]

T5C — Electronic principles: capacitance; inductance; current flow in circuits; alternating current; definition of RF; definition of polarity; DC power calculations; impedance

T5C01   What is the ability to store energy in an electric field called?

A. Inductance
B. Resistance
C. Tolerance
D. Capacitance

D   Storing energy in an electric field is called *capacitance* and it is measured in farads (F). [*Ham Radio License Manual*, page 3-9]

T5C02   What is the basic unit of capacitance?

A. The farad
B. The ohm
C. The volt
D. The henry

A   See question T5C01. [*Ham Radio License Manual*, page 3-9]

T5C03   What is the ability to store energy in a magnetic field called?

A. Admittance
B. Capacitance
C. Resistance
D. Inductance

D   Storing energy in a magnetic field is called *inductance* and it is measured in henrys (H). [*Ham Radio License Manual*, page 3-9]

T5C04   What is the basic unit of inductance?

A. The coulomb
B. The farad
C. The henry
D. The ohm

C   See question T5C03. [*Ham Radio License Manual*, page 3-9]

T5C05   What is the unit of frequency?

A. Hertz
B. Henry
C. Farad
D. Tesla

A   See question T5A12. [*Ham Radio License Manual*, page 2-3]

T5C06   **What does the abbreviation RF refer to?**

A. Radio frequency signals of all types
B. The resonant frequency of a tuned circuit
C. The real frequency transmitted as opposed to the apparent frequency
D. Reflective force in antenna transmission lines

A   If connected to a speaker, signals below 20 kHz produce sound waves that humans can hear, so we call them *audio frequency* or *AF* signals. Signals that have a frequency greater than 20,000 Hz (or 20 kHz) are *radio frequency* or *RF* signals. [*Ham Radio License Manual*, page 2-4]

T5C07   **A radio wave is made up of what type of energy?**

A. Pressure
B. Electromagnetic
C. Gravity
D. Thermal

B   See question T3B03. [*Ham Radio License Manual*, page 4-6]

T5C08   **What is the formula used to calculate electrical power in a DC circuit?**

A. Power (P) equals voltage (E) multiplied by current (I)
B. Power (P) equals voltage (E) divided by current (I)
C. Power (P) equals voltage (E) minus current (I)
D. Power (P) equals voltage (E) plus current (I)

A   In a dc circuit, power is calculated as the product of voltage and current ($P = E \times I$). See question T5D01. [*Ham Radio License Manual*, page 3-7]

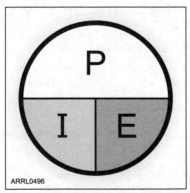

ARRL0496

This simple diagram will help you remember the power relationships. If you know any two of the quantities, you can find the third by covering up the unknown quantity. The positions of the remaining two symbols show if you have to multiply (side-by-side) or divide (one over the other).

T5C09   How much power is being used in a circuit when the applied voltage is 13.8 volts DC and the current is 10 amperes?

A. 138 watts
B. 0.7 watts
C. 23.8 watts
D. 3.8 watts

A $P = E \times I = 13.8 \text{ V} \times 10 \text{ A} = 138 \text{ W}$ [*Ham Radio License Manual*, page 3-7]

T5C10   How much power is being used in a circuit when the applied voltage is 12 volts DC and the current is 2.5 amperes?

A. 4.8 watts
B. 30 watts
C. 14.5 watts
D. 0.208 watts

B $P = E \times I = 12 \text{ V} \times 2.5 \text{ A} = 30 \text{ W}$ [*Ham Radio License Manual*, page 3-7]

T5C11   How many amperes are flowing in a circuit when the applied voltage is 12 volts DC and the load is 120 watts?

A. 0.1 amperes
B. 10 amperes
C. 12 amperes
D. 132 amperes

B $I = P / E = 120 \text{ W} / 12 \text{ V} = 10 \text{ A}$ [*Ham Radio License Manual*, page 3-7]

T5C12   What is impedance?

A. A measure of the opposition to AC current flow in a circuit
B. The inverse of resistance
C. The Q or Quality Factor of a component
D. The power handling capability of a component

A   The combination of resistance and reactance is called *impedance*, represented by the capital letter Z, and is also measured in ohms (Ω). *Impedance* is often used as a general term to mean the circuit's opposition to ac current flow. [*Ham Radio License Manual*, page 3-10]

T5C13   What is a unit of impedance?

A. Volts
B. Amperes
C. Coulombs
D. Ohms

D   See question T5C12. [*Ham Radio License Manual*, page 3-10]

T5C14 **What is the proper abbreviation for megahertz?**

A. mHz
B. mhZ
C. Mhz
D. MHz

D See the table of metric prefixes at the beginning of section T5B. [*Ham Radio License Manual*, page 2-3]

T5D — Ohm's Law: formulas and usage; components in series and parallel

T5D01 **What formula is used to calculate current in a circuit?**

A. Current (I) equals voltage (E) multiplied by resistance (R)
B. Current (I) equals voltage (E) divided by resistance (R)
C. Current (I) equals voltage (E) added to resistance (R)
D. Current (I) equals voltage (E) minus resistance (R)

B

$I = E / R$

$E = I \times R$

$R = E / I$

[*Ham Radio License Manual*, page 3-5]

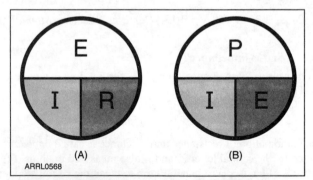

(A) (B)

ARRL0568

These simple diagrams will help you remember the Ohm's Law (A) and power (B) relationships. If you know any two of the quantities, the equation to find the third is shown by covering up the unknown quantity. The positions of the remaining two symbols show if you have to multiply (side-by-side) or divide (one above the other).

T5D02   **What formula is used to calculate voltage in a circuit?**

A. Voltage (E) equals current (I) multiplied by resistance (R)
B. Voltage (E) equals current (I) divided by resistance (R)
C. Voltage (E) equals current (I) added to resistance (R)
D. Voltage (E) equals current (I) minus resistance (R)

A   See question T5D01. [*Ham Radio License Manual*, page 3-5]

T5D03   **What formula is used to calculate resistance in a circuit?**

A. Resistance (R) equals voltage (E) multiplied by current (I)
B. Resistance (R) equals voltage (E) divided by current (I)
C. Resistance (R) equals voltage (E) added to current (I)
D. Resistance (R) equals voltage (E) minus current (I)

B   See question T5D01. [*Ham Radio License Manual*, page 3-5]

T5D04   **What is the resistance of a circuit in which a current of 3 amperes flows through a resistor connected to 90 volts?**

A. 3 ohms
B. 30 ohms
C. 93 ohms
D. 270 ohms

B $R = E / I = 90 \text{ V} / 3 \text{ A} = 30 \ \Omega$ [*Ham Radio License Manual*, page 3-6]

T5D05   **What is the resistance in a circuit for which the applied voltage is 12 volts and the current flow is 1.5 amperes?**

A. 18 ohms
B. 0.125 ohms
C. 8 ohms
D. 13.5 ohms

C $R = E / I = 12 \text{ V} / 1.5 \text{ A} = 8 \ \Omega$ [*Ham Radio License Manual*, page 3-6]

T5D06   **What is the resistance of a circuit that draws 4 amperes from a 12-volt source?**

A. 3 ohms
B. 16 ohms
C. 48 ohms
D. 8 ohms

A $R = E / I = 12 \text{ V} / 4 \text{ A} = 3 \ \Omega$ [*Ham Radio License Manual*, page 3-6]

T5D07   What is the current in a circuit with an applied voltage of 120 volts and a resistance of 80 ohms?

A. 9600 amperes
B. 200 amperes
C. 0.667 amperes
D. 1.5 amperes

D $I = E / R = 120$ V $/ 80$ $\Omega = 1.5$ A [*Ham Radio License Manual*, page 3-6]

T5D08   What is the current through a 100-ohm resistor connected across 200 volts?

A. 20,000 amperes
B. 0.5 amperes
C. 2 amperes
D. 100 amperes

C $I = E / R = 200$ V $/ 100$ $\Omega = 2$ A [*Ham Radio License Manual*, page 3-6]

T5D09   What is the current through a 24-ohm resistor connected across 240 volts?

A. 24,000 amperes
B. 0.1 amperes
C. 10 amperes
D. 216 amperes

C $I = E / R = 240$ V $/ 24$ $\Omega = 10$ A [*Ham Radio License Manual*, page 3-6]

T5D10   What is the voltage across a 2-ohm resistor if a current of 0.5 amperes flows through it?

A. 1 volt
B. 0.25 volts
C. 2.5 volts
D. 1.5 volts

A $E = I \times R = 0.5$ A $\times 2$ $\Omega = 1$ V [*Ham Radio License Manual*, page 3-6]

T5D11   What is the voltage across a 10-ohm resistor if a current of 1 ampere flows through it?

A. 1 volt
B. 10 volts
C. 11 volts
D. 9 volts

B $E = I \times R = 1 \times 10 = 10$ V [*Ham Radio License Manual*, page 3-7]

T5D12   What is the voltage across a 10-ohm resistor if a current of 2 amperes flows through it?

A. 8 volts
B. 0.2 volts
C. 12 volts
D. 20 volts

D $E = I \times R = 2 \times 10 = 20$ V [*Ham Radio License Manual*, page 3-7]

T5D13   What happens to current at the junction of two components in series?

A. It divides equally between them
B. It is unchanged
C. It divides based on the on the value of the components
D. The current in the second component is zero

B   At the junction of two components in a series circuit, the same current flows in each component. See question T5A13. [*Ham Radio License Manual*, page 3-2]

T5D14   What happens to current at the junction of two components in parallel?

A. It divides between them dependent on the value of the components
B. It is the same in both components
C. Its value doubles
D. Its value is halved

A   If two components are connected in parallel, the current divides between them. How much goes through each component depends on characteristics or value of the component. See question T5A13. [*Ham Radio License Manual*, page 3-2]

T5D15   What is the voltage across each of two components in series with a voltage source?

A. The same voltage as the source
B. Half the source voltage
C. It is determined by the type and value of the components
D. Twice the source voltage

D   When components are connected in series with a source of voltage the voltage divides between the components, depending on their type and value. [*Ham Radio License Manual*, page 3-3]

T5D16 **What is the voltage across each of two components in parallel with a voltage source?**

A. It is determined by the type and value of the components
B. Half the source voltage
C. Twice the source voltage
D. The same voltage as the source

C When components are connected in parallel with a source of voltage the voltage across each is the same as that of the source. See question T5A13. [*Ham Radio License Manual*, page 3-2]

Electrical Components

SUBELEMENT T6 — Electrical components; circuit diagrams; component functions
[4 exam questions — 4 groups]

T6A — Electrical components: fixed and variable resistors; capacitors and inductors; fuses; switches; batteries

T6A01   What electrical component opposes the flow of current in a DC circuit?

A. Inductor
B. Resistor
C. Voltmeter
D. Transformer

B   *Resistors* oppose the flow of electrical current in an ac or dc circuit. [*Ham Radio License Manual*, page 3-9]

T6A02   What type of component is often used as an adjustable volume control?

A. Fixed resistor
B. Power resistor
C. Potentiometer
D. Transformer

C   A variable resistor, also called a *potentiometer* (poh-ten-chee-AH-meh-tur) or *pot*, is frequently used to adjust voltage or potential, such as for a volume control. [*Ham Radio License Manual*, page 3-9]

T6A03   What electrical parameter is controlled by a potentiometer?

A. Inductance
B. Resistance
C. Capacitance
D. Field strength

B   See question T6A02. [*Ham Radio License Manual*, page 3-9]

T6A04   **What electrical component stores energy in an electric field?**

A. Resistor
B. Capacitor
C. Inductor
D. Diode

B   *Capacitors* store electrical energy in the *electric field* created by a voltage between two conducting surfaces or *electrodes* that are separated by an insulator called a *dielectric*. [*Ham Radio License Manual*, page 3-9]

T6A05   **What type of electrical component consists of two or more conductive surfaces separated by an insulator?**

A. Resistor
B. Potentiometer
C. Oscillator
D. Capacitor

D   See question T6A04. [*Ham Radio License Manual*, page 3-9]

T6A06   **What type of electrical component stores energy in a magnetic field?**

A. Resistor
B. Capacitor
C. Inductor
D. Diode

C   *Inductors* store energy in the *magnetic field* created by current flowing in a wire. Inductors are sometimes wound around a *core* of magnetic material that concentrates the magnetic energy. [*Ham Radio License Manual*, page 3-9]

T6A07   **What electrical component usually is constructed as a coil of wire?**

A. Switch
B. Capacitor
C. Diode
D. Inductor

D   See question T6A06. [*Ham Radio License Manual*, page 3-9]

T6A08   **What electrical component is used to connect or disconnect electrical circuits?**

A. Magnetron
B. Switch
C. Thermistor
D. All of these choices are correct

B   *Switches* and *relays* control current through a circuit by connecting and disconnecting paths for current to follow. [*Ham Radio License Manual*, page 3-13]

T6A09   What electrical component is used to protect other circuit components from current overloads?

A. Fuse
B. Capacitor
C. Inductor
D. All of these choices are correct

A   *Fuses* interrupt current overloads by melting a short length of metal. [*Ham Radio License Manual*, page 3-12]

T6A10   Which of the following battery types is rechargeable?

A. Nickel-metal hydride
B. Lithium-ion
C. Lead-acid gel-cell
D. All of these choices are correct

D   [*Ham Radio License Manual*, page 5-17]

Battery Types and Characteristics

Battery Style	Chemistry Type	Fully-Charged Voltage	Energy Rating (average)
AAA	Alkaline — Disposable	1.5 V	1100 mAh
AA	Alkaline — Disposable	1.5 V	2600 – 3200 mAh
AA	Carbon-Zinc — Disposable	1.5 V	600 mAh
AA	Nickel-Cadmium (NiCd) — Rechargeable	1.2 V	700 mAh
AA	Nickel-Metal Hydride (NiMH) — Rechargeable	1.2 V	1500 – 2200 mAh
C	Alkaline — Disposable	1.5 V	7500 mAh
D	Alkaline — Disposable	1.5 V	14,000 mAh
9 V	Alkaline — Disposable	9 V	580 mAh
9 V	Nickel-Cadmium (NiCd) — Rechargeable	9 V	110 mAh
9 V	Nickel-Metal Hydride — Rechargeable	9 V	150 mAh
Coin Cells	Lithium — Disposable	3 – 3.3 V	25 – 1000 mAh
Packs	Lithium ion (Li-ion) — Rechargeable	3.3 – 3.6 V per cell	Varies
Storage	Lead-acid — Rechargeable	2 V per cell	Varies

T6A11   Which of the following battery types is not rechargeable?

A. Nickel-cadmium
B. Carbon-zinc
C. Lead-acid
D. Lithium-ion

B   See the table for question T6A10. [*Ham Radio License Manual*, page 5-17]

T6B — Semiconductors: basic principles and applications of solid state devices; diodes and transistors

T6B01   What class of electronic components uses a voltage or current signal to control current flow?

A. Capacitors
B. Inductors
C. Resistors
D. Transistors

D   *Transistors* are components made of N- and P-type semiconductor patterns. The transistor's electrodes are contacts made to a certain piece of the pattern. Transistors use small voltages and currents to control larger ones. [*Ham Radio License Manual*, page 3-12]

T6B02   What electronic component allows current to flow in only one direction?

A. Resistor
B. Fuse
C. Diode
D. Driven element

C   A semiconductor that only allows current flow in one direction is called a *diode*. A diode has two electrodes, an *anode* and a *cathode*. On a diode the cathode is usually identified by a stripe marked on the component. Heavy-duty diodes that can handle large voltages and currents are called *rectifiers*. [*Ham Radio License Manual*, page 3-12]

T6B03   Which of these components can be used as an electronic switch or amplifier?

A. Oscillator
B. Potentiometer
C. Transistor
D. Voltmeter

C   With the appropriate external circuit and a source of power, transistors can amplify or switch voltages and currents. Using small signals to control or amplify larger signals is called *gain*. [*Ham Radio License Manual*, page 3-12]

T6B04   Which of the following components can consist of three layers of semiconductor material?

A. Alternator
B. Transistor
C. Triode
D. Pentagrid converter

B   A transistor is made from up to three layers of P- and N-type semiconductor material. [*Ham Radio License Manual*, page 3-12]

T6B05 Which of the following electronic components can amplify signals?

A. Transistor
B. Variable resistor
C. Electrolytic capacitor
D. Multi-cell battery

A See question T6B03. [*Ham Radio License Manual*, page 3-12]

T6B06 How is the cathode lead of a semiconductor diode often marked on the package?

A. With the word cathode
B. With a stripe
C. With the letter C
D. With the letter K

B See question T6B02. [*Ham Radio License Manual*, page 3-12]

T6B07 What does the abbreviation LED stand for?

A. Low Emission Diode
B. Light Emitting Diode
C. Liquid Emission Detector
D. Long Echo Delay

B A special type of diode, the light-emitting diode or LED, gives off light when current flows through it. The material from which the LED is made determines the color of light emitted. LEDs are most often used as visual indicators. [*Ham Radio License Manual*, page 3-12]

T6B08 What does the abbreviation FET stand for?

A. Field Effect Transistor
B. Fast Electron Transistor
C. Free Electron Transmitter
D. Frequency Emission Transmitter

A There are two common types of transistors: *bipolar junction transistors (BJT)* and *field-effect transistors (FET)*. [*Ham Radio License Manual*, page 3-12]

T6B09 What are the names of the two electrodes of a diode?

A. Plus and minus
B. Source and drain
C. Anode and cathode
D. Gate and base

C See question T6B02. [*Ham Radio License Manual*, page 3-12]

T6B10 Which of the following could be the primary gain-producing component in an RF power amplifier?

A. Transformer
B. Transistor
C. Reactor
D. Resistor

B RF power transistors are used as the primary gain-producing component in RF power amplifiers. [*Ham Radio License Manual*, page 3-12]

T6B11 What is the term that describes a device's ability to amplify a signal?

A. Gain
B. Forward resistance
C. Forward voltage drop
D. On resistance

A See question T6B03. [*Ham Radio License Manual*, page 3-12]

T6C — Circuit diagrams; schematic symbols

T6C01 What is the name of an electrical wiring diagram that uses standard component symbols?

A. Bill of materials
B. Connector pinout
C. Schematic
D. Flow chart

C *Schematics* are a visual description of a circuit and its components that use standardized drawings called *circuit symbols*. [*Ham Radio License Manual*, page 3-14]

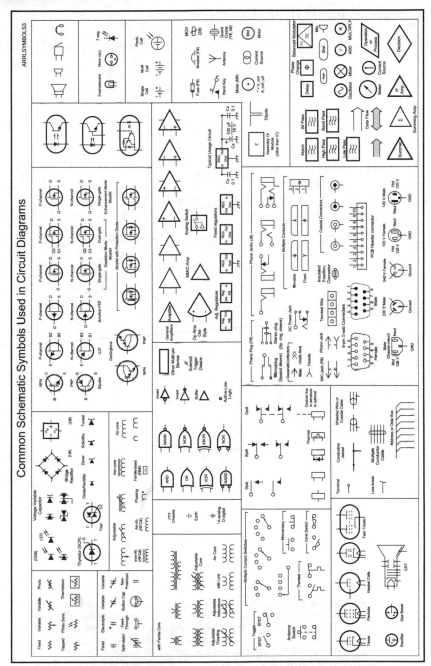

Common Schematic Symbols Used in Circuit Diagrams

ARRLSYMBOLS3

Use this symbol guide for questions T6C02 to T6C11, T6D03, and T6D10.
Symbols are used when drawing a circuit because there are so many types of
components. Radio and electrical designers use them as a convenient way of
describing a circuit.

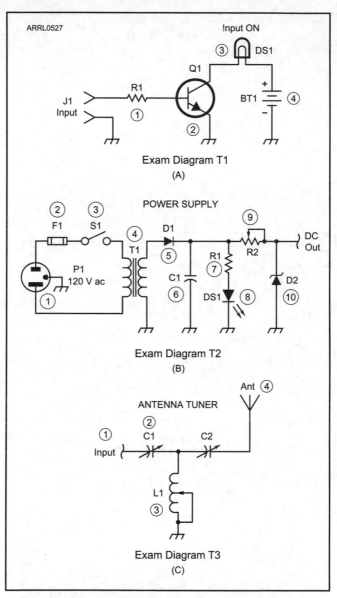

ARRL0527

Input ON

③ DS1

Q1

R1
①

J1
Input

BT1 ④
+
–

Exam Diagram T1
(A)

POWER SUPPLY

② ③
F1 S1

④
T1

D1

⑤

⑨

R2

DC
Out

P1
120 V ac

①

R1
⑦

C1
⑥

DS1 ⑧

D2
⑩

Exam Diagram T2
(B)

Ant ④

ANTENNA TUNER

②
C1

C2

①
Input

L1
③

Exam Diagram T3
(C)

A schematic diagram describes complex circuits using symbols representing each type of component. Lines and dots show electrical connections between the components, but may not correspond to actual wires. Use diagrams T1 (A), T2 (B) and T3 (C) that are used on the Technician exam for questions T6C02 through T6C11, T6D03, and T6D10.

T6C02   What is component 1 in figure T1?

A. Resistor
B. Transistor
C. Battery
D. Connector

A   For questions T6C02 through T6C11, T6D03, and T6D10, refer to schematic diagrams T1, T2, and T3. These are the schematic diagrams used on the license exam. [*Ham Radio License Manual,* page 3-16]

T6C03   What is component 2 in figure T1?

A. Resistor
B. Transistor
C. Indicator lamp
D. Connector

B   [*Ham Radio License Manual*, page 3-16]

T6C04   What is component 3 in figure T1?

A. Resistor
B. Transistor
C. Lamp
D. Ground symbol

C   [*Ham Radio License Manual*, page 3-16]

T6C05   What is component 4 in figure T1?

A. Resistor
B. Transistor
C. Battery
D. Ground symbol

C   [*Ham Radio License Manual*, page 3-16]

T6C06   What is component 6 in figure T2?

A. Resistor
B. Capacitor
C. Regulator IC
D. Transistor

B   [*Ham Radio License Manual*, page 3-16]

T6C07   What is component 8 in figure T2?

A. Resistor
B. Inductor
C. Regulator IC
D. Light emitting diode

D   [*Ham Radio License Manual*, page 3-16]

T6C08   What is component 9 in figure T2?

A. Variable capacitor
B. Variable inductor
C. Variable resistor
D. Variable transformer

C   [*Ham Radio License Manual*, page 3-16]

T6C09   What is component 4 in figure T2?

A. Variable inductor
B. Double-pole switch
C. Potentiometer
D. Transformer

D   [*Ham Radio License Manual*, page 3-16]

T6C10   What is component 3 in figure T3?

A. Connector
B. Meter
C. Variable capacitor
D. Variable inductor

D   [*Ham Radio License Manual*, page 3-16]

T6C11   What is component 4 in figure T3?

A. Antenna
B. Transmitter
C. Dummy load
D. Ground

A   [*Ham Radio License Manual*, page 3-16]

T6C12   What do the symbols on an electrical schematic represent?

A. Electrical components
B. Logic states
C. Digital codes
D. Traffic nodes

A   See question T6C01. [*Ham Radio License Manual*, page 3-14]

T6C13   Which of the following is accurately represented in electrical schematics?

A. Wire lengths
B. Physical appearance of components
C. The way components are interconnected
D. All of these choices are correct

C   See question T6C02. [*Ham Radio License Manual, page* 3-14]

T6D — Component functions: rectification; switches; indicators; power supply components; resonant circuit; shielding; power transformers; integrated circuits

T6D01   Which of the following devices or circuits changes an alternating current into a varying direct current signal?

A. Transformer
B. Rectifier
C. Amplifier
D. Reflector

B   If an ac voltage is applied to a diode, the result is a pulsing dc current. [*Ham Radio License Manual*, page 3-12]

T6D02   What is a relay?

A. An electrically-controlled switch
B. A current controlled amplifier
C. An optical sensor
D. A pass transistor

A   A switch is operated manually while a relay is a switch controlled by an electromagnet. [*Ham Radio License Manual*, page 3-13]

T6D03   What type of switch is represented by component 3 in figure T2?

A. Single-pole single-throw
B. Single-pole double-throw
C. Double-pole single-throw
D. Double-pole double-throw

A   A double-throw (DT) switch can route current through either of two paths while a single-throw (ST) switch can only open or close a single path. [*Ham Radio License Manual*, page 3-14]

T6D04   Which of the following displays an electrical quantity as a numeric value?

A. Potentiometer
B. Transistor
C. Meter
D. Relay

C   An *indicator* is either ON or OFF, such as a power indicator or a label that appears when you are transmitting. A *meter* provides information as a value in the form of numbers or on a numeric scale. A *display* combines indicators, numbers, and labels. A *liquid crystal display* or LCD is used on the front panel of many radios and test instruments. [*Ham Radio License Manual*, page 3-14]

T6D05   What type of circuit controls the amount of voltage from a power supply?

A. Regulator
B. Oscillator
C. Filter
D. Phase inverter

A   A *regulated supply* uses a *regulator* circuit to minimize the amount of voltage change at different current levels. [*Ham Radio License Manual*, page 5-16]

T6D06 What component is commonly used to change 120V AC house current to a lower AC voltage for other uses?

A. Variable capacitor
B. Transformer
C. Transistor
D. Diode

B *Transformers* are made from two or more inductors that share their stored energy. This allows energy to be transferred from one inductor to another while changing the combination of voltage and current. [*Ham Radio License Manual*, page 3-9]

T6D07 Which of the following is commonly used as a visual indicator?

A. LED
B. FET
C. Zener diode
D. Bipolar transistor

A See question T6B07. [*Ham Radio License Manual*, page 3-12]

T6D08 Which of the following is combined with an inductor to make a tuned circuit?

A. Resistor
B. Zener diode
C. Potentiometer
D. Capacitor

D Circuits that contain both a capacitor and an inductor are called *resonant circuits* or *tuned circuits*. [*Ham Radio License Manual*, page 3-10]

T6D09 What is the name of a device that combines several semiconductors and other components into one package?

A. Transducer
B. Multi-pole relay
C. Integrated circuit
D. Transformer

C An *integrated circuit* (*IC* or *chip*) is made of many components connected together as a useful circuit and packaged as a single component. [*Ham Radio License Manual*, page 3-12]

T6D10   What is the function of component 2 in Figure T1?
A. Give off light when current flows through it
B. Supply electrical energy
C. Control the flow of current
D. Convert electrical energy into radio waves

C   See question T6B01. [*Ham Radio License Manual*, page 3-12]

T6D11   Which of the following is a resonant or tuned circuit?
A. An inductor and a capacitor connected in series or parallel to form a filter
B. A type of voltage regulator
C. A resistor circuit used for reducing standing wave ratio
D. A circuit designed to provide high-fidelity audio

A   A tuned circuit acts as a filter, passing or rejecting signals at its resonant frequency. [*Ham Radio License Manual*, page 3-10]

T6D12   Which of the following is a common reason to use shielded wire?
A. To decrease the resistance of DC power connections
B. To increase the current carrying capability of the wire
C. To prevent coupling of unwanted signals to or from the wire
D. To couple the wire to other signals

C   The use of shielded wire and shielded cables prevents coupling with unwanted signals and undesired radiation. [*Ham Radio License Manual, page 9-9]*

Station Equipment

SUBELEMENT T7 — Station equipment: common transmitter and receiver problems; antenna measurements; troubleshooting; basic repair and testing [4 exam questions — 4 groups]

T7A — Station equipment: receivers; transmitters; transceivers; modulation; transverters; transmit and receive amplifiers

> **T7A01** Which term describes the ability of a receiver to detect the presence of a signal?
>
> A. Linearity
> B. Sensitivity
> C. Selectivity
> D. Total Harmonic Distortion

B A receiver's *sensitivity* determines its ability to detect signals. Higher sensitivity means a receiver can detect weaker signals. [*Ham Radio License Manual*, page 5-9]

> **T7A02** What is a transceiver?
>
> A. A type of antenna switch
> B. A unit combining the functions of a transmitter and a receiver
> C. A component in a repeater that filters out unwanted interference
> D. A type of antenna matching network

B Most amateur equipment combines the transmitter and receiver into a single piece of equipment called a *transceiver* (abbreviated XCVR). A *transmit-receive (TR) switch* allows the transmitter and receiver to share a single antenna. [*Ham Radio License Manual,* page 2-7]

A basic amateur station is made up of a transmitter and receiver connected to an antenna with a feed line. The transmit-receive (TR) switch allows the transmitter and receiver to share the antenna. A transceiver combines the transmitter, receiver, and TR switch in a single package.

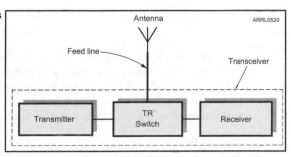

T7A03 Which of the following is used to convert a radio signal from one frequency to another?

A. Phase splitter
B. Mixer
C. Inverter
D. Amplifier

B *Mixers* combine two RF signals and shift one of them to a third frequency. [*Ham Radio License Manual*, page 3-18]

T7A04 Which term describes the ability of a receiver to discriminate between multiple signals?

A. Discrimination ratio
B. Sensitivity
C. Selectivity
D. Harmonic distortion

C *Selectivity* is the ability of a receiver to discriminate between signals, retrieving only the information from the desired signal in the presence of unwanted signals. High selectivity means that a receiver can operate properly even in the presence of strong signals on nearby frequencies. [*Ham Radio License Manual*, page 5-9]

T7A05 What is the name of a circuit that generates a signal at a | specific frequency?

A. Reactance modulator
B. Product detector
C. Low-pass filter
D. Oscillator

D An *oscillator* produces a steady signal at one frequency. [*Ham Radio License Manual*, page 3-17]

T7A06 What device converts the RF input and output of a transceiver to another band?

A. High-pass filter
B. Low-pass filter
C. Transverter
D. Phase converter

C A *transverter* allows a transceiver to be used on one or more new bands. Low-power transmitter output signals are shifted to the new output frequency where they are amplified for transmission. A *receiving converter* mixer shifts input signals to the desired band where they are received as regular signals by the transceiver. [*Ham Radio License Manual*, page 5-11]

T7A07 **What is meant by PTT?**

A. Pre-transmission tuning to reduce transmitter harmonic emission
B. Precise tone transmissions used to limit repeater access to only certain signals
C. A primary transformer tuner use to match antennas
D. The push-to-talk function that switches between receive and transmit

D Switching between receive and transmit on voice can be performed manually with a *push-to-talk* (PTT) switch or an automatic *voice-operated transmitter* control circuit (VOX). [*Ham Radio License Manual*, page 5-7]

T7A08 **Which of the following describes combining speech with an RF carrier signal?**

A. Impedance matching
B. Oscillation
C. Modulation
D. Low-pass filtering

C The process of combining data or voice signals with an RF signal is *modulation*. A circuit that performs the modulation function is therefore called a *modulator*. [*Ham Radio License Manual*, page 3-17]

T7A09 **What is the function of the SSB/CW-FM switch on a VHF power amplifier?**

A. Change the mode of the transmitted signal
B. Set the amplifier for proper operation in the selected mode
C. Change the frequency range of the amplifier to operate in the proper portion of the band
D. Reduce the received signal noise

B An RF power amplifier is used to increase the output power from low-power handheld transceivers and medium-power base or mobile rigs. Commercial amplifiers used on VHF can be used with SSB, FM, or CW signals but need to be set to the right mode to operate properly. This is controlled by a switch on the amplifier's front panel that changes the amplifier from the CW and FM setting to the SSB setting. [*Ham Radio License Manual*, page 5-10]

T7A10 **What device increases the low-power output from a handheld transceiver?**

A. A voltage divider
B. An RF power amplifier
C. An impedance network
D. All of these choices are correct

B See question T7A09. [*Ham Radio License Manual*, page 5-10]

T7A11   Where is an RF preamplifier installed?

A. Between the antenna and receiver
B. At the output of the transmitter's power amplifier
C. Between a transmitter and antenna tuner
D. At the receiver's audio output

A   If a receiver is not sensitive enough, a *preamplifier* or "preamp" can be used. The preamp is connected between the antenna and receiver. [*Ham Radio License Manual*, page 5-9]

T7B — Common transmitter and receiver problems: symptoms of overload and overdrive; distortion; causes of interference; interference and consumer electronics; part 15 devices; over-modulation; RF feedback; off frequency signals

T7B01   What can you do if you are told your FM handheld or mobile transceiver is over-deviating?

A. Talk louder into the microphone
B. Let the transceiver cool off
C. Change to a higher power level
D. Talk farther away from the microphone

D   An overmodulated FM signal has excessive deviation and is said to be *overdeviating*. Overdeviation is usually caused by speaking too loudly into the microphone and may cause interference on adjacent channels. It often generates noise or distortion on voice peaks, called "breaking up." To reduce overdeviation, speak more softly or move the microphone farther from your mouth. [*Ham Radio License Manual*, page 5-8]

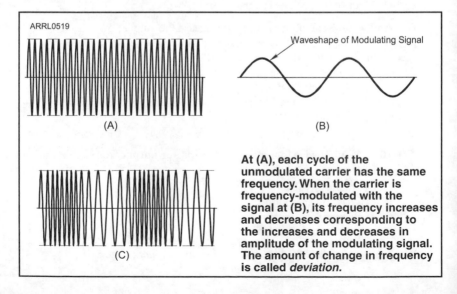

ARRL0519

(A)

Waveshape of Modulating Signal

(B)

(C)

At (A), each cycle of the unmodulated carrier has the same frequency. When the carrier is frequency-modulated with the signal at (B), its frequency increases and decreases corresponding to the increases and decreases in amplitude of the modulating signal. The amount of change in frequency is called *deviation*.

T7B02   What would cause a broadcast AM or FM radio to receive an amateur radio transmission unintentionally?

A. The receiver is unable to reject strong signals outside the AM or FM band
B. The microphone gain of the transmitter is turned up too high
C. The audio amplifier of the transmitter is overloaded
D. The deviation of an FM transmitter is set too low

A   See also question T7B03. [*Ham Radio License Manual*, page 9-8]

T7B03   Which of the following can cause radio frequency interference?

A. Fundamental overload
B. Harmonics
C. Spurious emissions
D. All of these choices are correct

D   Very strong signals may overwhelm a receiver's ability to reject them. This is called *fundamental overload*. Overload usually results in severe interference to all channels of a broadcast TV, AM, or FM receiver. A *high-pass filter* can be connected at the antenna input of FM and TV receivers to reject strong lower-frequency signals from amateur HF transmitters. A *band-reject* or *notch filter* can be used to reduce interference from amateur VHF or UHF signals. *Harmonics* of the desired output signal and other *spurious emissions* that can cause interference to nearby equipment. To prevent harmonics from being radiated by your station, a low-pass or band-pass filter must be installed at the transmitter's connection to the antenna feed line. [*Ham Radio License Manual*, page 9-8]

T7B04   Which of the following is a way to reduce or eliminate interference from an amateur transmitter to a nearby telephone?

A. Put a filter on the amateur transmitter
B. Reduce the microphone gain
C. Reduce the SWR on the transmitter transmission line
D. Put an RF filter on the telephone

D   Low-pass RF filters connected at the modular jack are the best way to reduce RFI to telephones by blocking RF signals on the telephone line. [*Ham Radio License Manual*, page 9-8]

T7B05 How can overload of a non-amateur radio or TV receiver by an amateur signal be reduced or eliminated?

A. Block the amateur signal with a filter at the antenna input of the affected receiver
B. Block the interfering signal with a filter on the amateur transmitter
C. Switch the transmitter from FM to SSB
D. Switch the transmitter to a narrow-band mode

A   See question T7B03. [*Ham Radio License Manual*, page 9-9]

T7B06 Which of the following actions should you take if a neighbor tells you that your station's transmissions are interfering with their radio or TV reception?

A. Make sure that your station is functioning properly and that it does not cause interference to your own radio or television when it is tuned to the same channel
B. Immediately turn off your transmitter and contact the nearest FCC office for assistance
C. Tell them that your license gives you the right to transmit and nothing can be done to reduce the interference
D. Install a harmonic doubler on the output of your transmitter and tune it until the interference is eliminated

A   Regardless of the source, you can reduce or eliminate much interference by making sure your own station follows good amateur practices for grounding and filtering. Eliminate interference to your own home appliances and televisions first. [*Ham Radio License Manual*, page 9-9]

T7B07 Which of the following can reduce overload to a VHF transceiver from a nearby FM broadcast station?

A. RF preamplifier
B. Double-shielded coaxial cable
C. Using headphones instead of the speaker
D. Band-reject filter

D   See question T7B03. [*Ham Radio License Manual*, page 9-9]

T7B08   What should you do if something in a neighbor's home is causing harmful interference to your amateur station?

A. Work with your neighbor to identify the offending device
B. Politely inform your neighbor about the rules that prohibit the use of devices that cause interference
C. Check your station and make sure it meets the standards of good amateur practice
D. All of these choices are correct

D   There are several simple steps to take:

• Make sure your station meets the standards of good amateur practices.

• You can offer to help determine the source of interference. Severe noise often indicates defective equipment that could be a safety or fire hazard.

• You may have to politely explain to the neighbor that FCC rules prohibit them from using a device that causes harmful interference. [*Ham Radio License Manual*, page 9-10]

T7B09   What is a Part 15 device?

A. An unlicensed device that may emit low-powered radio signals on frequencies used by a licensed service
B. An amplifier that has been type-certified for amateur radio
C. A device for long-distance communications using special codes sanctioned by the International Amateur Radio Union
D. A type of test set used to determine whether a transmitter complies with FCC regulation 91.15

A   Unlicensed Part 15 devices may use low-power RF communications or radiate low-power signals on frequencies used by licensed services, such as Amateur Radio. [*Ham Radio License Manual*, page 9-10]

T7B10   What might be a problem if you receive a report that your audio signal through the repeater is distorted or unintelligible?

A. Your transmitter is slightly off frequency
B. Your batteries are running low
C. You are in a bad location
D. All of these choices are correct

D   Being slightly off frequency sometimes happens when a radio control knob or key gets bumped, changing frequency by a small amount. You could also be causing excessive deviation by speaking too loudly into the microphone. Either lower your voice or hold the microphone farther away from your mouth. You could be transmitting from a bad location where your signal is distorted at the repeater input. Weak or *low* batteries can also cause distorted audio. [*Ham Radio License Manual*, page 6-5]

T7B11 **What is a symptom of RF feedback in a transmitter or transceiver?**

A. Excessive SWR at the antenna connection
B. The transmitter will not stay on the desired frequency
C Reports of garbled, distorted, or unintelligible voice transmissions
D. Frequent blowing of power supply fuses

C RF current flowing in sensitive audio cables or data cables can interfere with your station's normal function.. Often called "RF feedback," this type of interference can cause distorted transmitted audio, for example. [*Ham Radio License Manual*, page 9-6]

T7B12 **What should be the first step to resolve cable TV interference from your ham radio transmission?**

A. Add a low-pass filter to the TV antenna input
B. Add a high-pass filter to the TV antenna input
C. Add a preamplifier to the TV antenna input
D. Be sure all TV coaxial connectors are installed properly

D A possible source of interference to hams and to cable TV customers, is *leakage*. Leakage can refer to amateur signals getting into the cable TV feed line or to cable signals getting out and being radiated. The most common cause of leakage in either direction is faulty coaxial connectors on the cable feed line that are either loose or improperly installed. [*Ham Radio License Manual*, page 9-9]

T7C — Antenna measurements and troubleshooting: measuring SWR; dummy loads; coaxial cables; causes of feed line failures

T7C01   What is the primary purpose of a dummy load?

A. To prevent transmitting signals over the air when making tests
B. To prevent over-modulation of a transmitter
C. To improve the efficiency of an antenna
D. To improve the signal-to-noise ratio of a receiver

A   To avoid interfering with other stations while you're adjusting your transmitter or measuring its output power, it's a good idea to use a *dummy load*. A dummy load is a heavy-duty resistor that can absorb and dissipate the output power from a transmitter. [*Ham Radio License Manual*, page 5-7]

T7C02   Which of the following instruments can be used to determine if an antenna is resonant at the desired operating frequency?

A. A VTVM
B. An antenna analyzer
C. A Q meter
D. A frequency counter

B   An *antenna analyzer* contains a very low-power signal source with an adjustable frequency and one or more meters to show the impedance and SWR. [*Ham Radio License Manual*, page 4-19]

T7C03   What, in general terms, is standing wave ratio (SWR)?

A. A measure of how well a load is matched to a transmission line
B. The ratio of high to low impedance in a feed line
C. The transmitter efficiency ratio
D. An indication of the quality of your station's ground connection

A   Because SWR is determined by the amounts of forward and reflected power, SWR in an antenna system is also a measure of the how well the antenna (or load) and feed line impedances are matched. [*Ham Radio License Manual*, page 4-11]

T7C04   What reading on an SWR meter indicates a perfect impedance match between the antenna and the feed line?

A. 2 to 1
B. 1 to 3
C. 1 to 1
D. 10 to 1

C   When there is no reflected power the SWR is 1:1. This condition is called a *perfect match*. [*Ham Radio License Manual*, page 4-11]

T7C05 **Why do most solid-state amateur radio transmitters reduce output power as SWR increases?**

A. To protect the output amplifier transistors
B. To comply with FCC rules on spectral purity
C. Because power supplies cannot supply enough current at high SWR
D. To improve the impedance match to the feed line

A Most amateur transmitting equipment is designed to work at full power with an SWR of 2:1 or lower. Higher SWR may cause the transmitter's protection circuits to reduce power to avoid damage to the output transistors. [*Ham Radio License Manual*, page 4-11]

T7C06 **What does an SWR reading of 4:1 indicate?**

A. Loss of -4 dB
B. Good impedance match
C. Gain of +4 dB
D. Impedance mismatch

D As more power is reflected, SWR increases. SWR greater than 1:1 is called an *impedance mismatch* or just *mismatch*. [*Ham Radio License Manual*, page 4-11]

T7C07 **What happens to power lost in a feed line?**

A. It increases the SWR
B. It comes back into your transmitter and could cause damage
C. It is converted into heat
D. It can cause distortion of your signal

C Power lost in a feed line is converted into and dissipated as heat. [*Ham Radio License Manual*, page 4-9]

T7C08 **What instrument other than an SWR meter could you use to determine if a feed line and antenna are properly matched?**

A. Voltmeter
B. Ohmmeter
C. Iambic pentameter
D. Directional wattmeter

D Many amateurs use a *directional wattmeter* that can read power flowing in either direction in the feed line. The forward and reflected power readings can then be converted the SWR by using a table or formula. [*Ham Radio License Manual*, page 4-19]

T7C09   Which of the following is the most common cause for failure of coaxial cables?

A. Moisture contamination
B. Gamma rays
C. The velocity factor exceeds 1.0
D. Overloading

A   Water in coaxial cable degrades the effectiveness of the braided shield and dramatically increases losses. Nicks, cuts and scrapes can all breach the jacket of coaxial cable allowing moisture contamination, the most common cause of coaxial cable failure. Prolonged exposure to the ultraviolet (UV) in sunlight will also cause the plastic in the jacket to degrade, causing small cracks that allow water into the cable. [*Ham Radio License Manual*, page 4-17]

T7C10   Why should the outer jacket of coaxial cable be resistant to ultraviolet light?

A. Ultraviolet resistant jackets prevent harmonic radiation
B. Ultraviolet light can increase losses in the cable's jacket
C. Ultraviolet and RF signals can mix, causing interference
D. Ultraviolet light can damage the jacket and allow water to enter the cable

D   See question T7C09. [*Ham Radio License Manual*, page 4-17]

T7C11   What is a disadvantage of air core coaxial cable when compared to foam or solid dielectric types?

A. It has more loss per foot
B. It cannot be used for VHF or UHF antennas
C. It requires special techniques to prevent water absorption
D. It cannot be used at below freezing temperatures

C   Connectors that can seal the coax core must be used with extra attention paid to waterproofing. Another technique is to pressurize the cable. [*Ham Radio License Manual*, page 4-18]

T7C12   What does a dummy load consist of?

A. A high-gain amplifier and a TR switch
B. A non-inductive resistor and a heat sink
C. A low-voltage power supply and a DC relay
D. A 50 ohm reactance used to terminate a transmission line

B   See question T7C01. [*Ham Radio License Manual*, page 5-7]

T7D — Basic repair and testing: soldering; using basic test instruments; connecting a voltmeter, ammeter, or ohmmeter

T7D01   Which instrument would you use to measure electric potential or electromotive force?

A. An ammeter
B. A voltmeter
C. A wavemeter
D. An ohmmeter

B   Voltage is measured with a *voltmeter*. [*Ham Radio License Manual*, |page 3-2]

T7D02   What is the correct way to connect a voltmeter to a circuit?

A. In series with the circuit
B. In parallel with the circuit
C. In quadrature with the circuit
D. In phase with the circuit

B   Voltmeters are connected in parallel with or "across" a component or circuit to measure voltage. See question T5A13. [*Ham Radio License Manual*, page 3-3]

T7D03   How is a simple ammeter connected to a circuit?

A. In series with the circuit
B. In parallel with the circuit
C. In quadrature with the circuit
D. In phase with the circuit

A   Ammeters are connected in series with a component or circuit to measure current. [*Ham Radio License Manual*, page 3-3]

T7D04   Which instrument is used to measure electric current?

A. An ohmmeter
B. A wavemeter
C. A voltmeter
D. An ammeter

D   An *ammeter is used to measure current*. [*Ham Radio License Manual*, page 3-1]

T7D05   What instrument is used to measure resistance?

A. An oscilloscope
B. A spectrum analyzer
C. A noise bridge
D. An ohmmeter

D   Resistance is measured with an *ohmmeter*. [*Ham Radio License Manual*, page 3-5]

T7D06   Which of the following might damage a multimeter?

A. Measuring a voltage too small for the chosen scale
B. Leaving the meter in the milliamps position overnight
C. Attempting to measure voltage when using the resistance setting
D. Not allowing it to warm up properly

C   Trying to measure voltage or connecting the probes to an energized circuit when the meter is set to measure resistance is a common way to damage a multimeter, for example. [*Ham Radio License Manual*, page 3-4]

T7D07   Which of the following measurements are commonly made using a multimeter?

A. SWR and RF power
B. Signal strength and noise
C. Impedance and reactance
D. Voltage and resistance

D   The multimeter measures all three primary electrical values of voltage, current, and resistance. [*Ham Radio License Manual*, page 3-4]

T7D08   Which of the following types of solder is best for radio and electronic use?

A. Acid-core solder
B. Silver solder
C. Rosin-core solder
D. Aluminum solder

C   Avoid acid-core solder which is corrosive to electronics. [*Ham Radio License Manual*, page 4-18]

T7D09   What is the characteristic appearance of a cold solder joint?
A. Dark black spots
B. A bright or shiny surface
C. A grainy or dull surface
D. A greenish tint

C   A grainy or dull surface indicates that the solder and the wires or terminals being soldered did not get hot enough at the same time. You should re-heat the joint and re-solder it. [*Ham Radio License Manual*, page 4-18]

T7D10   What is probably happening when an ohmmeter, connected across an unpowered circuit, initially indicates a low resistance and then shows increasing resistance with time?
A. The ohmmeter is defective
B. The circuit contains a large capacitor
C. The circuit contains a large inductor
D. The circuit is a relaxation oscillator

B   At first, when the capacitor is discharged, it performs like a low resistance to the meter. As voltage across the capacitor increases, less current flows, performing like a higher resistance. [*Ham Radio License Manual*, page 3-4]

T7D11   Which of the following precautions should be taken when measuring circuit resistance with an ohmmeter?
A. Ensure that the applied voltages are correct
B. Ensure that the circuit is not powered
C. Ensure that the circuit is grounded
D. Ensure that the circuit is operating at the correct frequency

B   See question T7D06. [*Ham Radio License Manual*, page 3-4]

T7D12   Which of the following precautions should be taken when measuring high voltages with a voltmeter?
A. Ensure that the voltmeter has very low impedance
B. Ensure that the voltmeter and leads are rated for use at the voltages to be measured
C. Ensure that the circuit is grounded through the voltmeter
D. Ensure that the voltmeter is set to the correct frequency

B   Look for the meter's voltage rating which is usually labeled on the face of the meter. Voltages beyond the meter's rating can "flashover" to other pieces of equipment or to you, creating a serious shock hazard. Ensure that the voltmeter and leads are rated for use at the voltages to be measured! [*Ham Radio License Manual*, page 3-4]

Modulation Modes

SUBELEMENT T8 — Modulation modes: amateur satellite operation; operating activities; non-voice and digital communications
[4 exam questions — 4 groups]

T8A — Modulation modes: bandwidth of various signals; choice of emission type

T8A01   Which of the following is a form of amplitude modulation?
A. Spread spectrum
B. Packet radio
C. Single sideband
D. Phase shift keying (PSK)

C   Single sideband or SSB is an amplitude modulation (AM) signal with the carrier and one sideband removed or *suppressed*. [*Ham Radio License Manual*, page 5-3]

T8A02   What type of modulation is most commonly used for VHF packet radio transmissions?
A. FM
B. SSB
C. AM
D. PSK

A   Frequency modulation or FM is commonly used for packet radio on VHF and UHF. The data modulates the FM signal as audio tones. [*Ham Radio License Manual*, page 5-4]

T8A03   Which type of voice mode is most often used for long-distance (weak signal) contacts on the VHF and UHF bands?

A. FM
B. DRM
C. SSB
D. PM

C   Because an SSB signal's power is concentrated into a narrow bandwidth, it is possible to communicate with SSB over much longer ranges and in poorer conditions than with FM or AM, particularly on the VHF and UHF bands. [*Ham Radio License Manual*, page 5-4]

T8A04   Which type of modulation is most commonly used for VHF and UHF voice repeaters?

A. AM
B. SSB
C. PSK
D. FM

D   Because of FM's excellent noise-rejection qualities, it the mode used by most VHF and UHF repeaters. [*Ham Radio License Manual*, page 5-4]

T8A05   Which of the following types of emission has the narrowest bandwidth?

A. FM voice
B. SSB voice
C. CW
D. Slow-scan TV

C   [*Ham Radio License Manual*, page 5-5]

Signal Bandwidths

Type of Signal	Typical Bandwidth
CW	150 Hz (0.15 kHz)
SSB digital	500 to 3000 Hz (0.5 to 3 kHz)
SSB voice	2 to 3 kHz
AM voice	6 kHz
AM broadcast	10 kHz
FM voice	10 to 15 kHz
FM broadcast	150 kHz
Fast-scan video broadcast	6 MHz

T8A06   Which sideband is normally used for 10 meter HF, VHF, and UHF single-sideband communications?

A. Upper sideband
B. Lower sideband
C. Suppressed sideband
D. Inverted sideband

A   Above 10 MHz, including all of the VHF and UHF bands, upper sideband (USB) is the standard choice. USB must also be used on the 60 meter channels. LSB is used for voice below 10 MHz. [*Ham Radio License Manual*, page 5-5]

T8A07   What is an advantage of single sideband (SSB) over FM for voice transmissions?

A. SSB signals are easier to tune
B. SSB signals are less susceptible to interference
C. SSB signals have narrower bandwidth
D. All of these choices are correct

C   See the table for question T8A03. [*Ham Radio License Manual*, page 5-4]

T8A08   What is the approximate bandwidth of a single sideband (SSB) voice signal?

A. 1 kHz
B. 3 kHz
C. 6 kHz
D. 15 kHz

B   See the table for question T8A05. [*Ham Radio License Manual*, page 5-5]

T8A09   What is the approximate bandwidth of a VHF repeater FM phone signal?

A. Less than 500 Hz
B. About 150 kHz
C. Between 10 and 15 kHz
D. Between 50 and 125 kHz

C   See the table for question T8A05. [*Ham Radio License Manual*, page 5-5]

T8A10   What is the typical bandwidth of analog fast-scan TV transmissions on the 70 centimeter band?

A. More than 10 MHz
B. About 6 MHz
C. About 3 MHz
D. About 1 MHz

B   See the table for question T8A05. [*Ham Radio License Manual*, page 5-5]

T8A11   What is the approximate maximum bandwidth required to transmit a CW signal?

A. 2.4 kHz
B. 150 Hz
C. 1000 Hz
D. 15 kHz

B   See the table for question T8A05. [*Ham Radio License Manual*, page 5-5]

T8B — Amateur satellite operation; Doppler shift; basic orbits; operating protocols; transmitter power considerations; telemetry and telecommand; satellite tracking

T8B01   What telemetry information is typically transmitted by satellite beacons?

A. The signal strength of received signals
B. Time of day accurate to plus or minus 1/10 second
C. Health and status of the satellite
D. All of these choices are correct

C   The telemetry data from a satellite contains information on the health and status of the satellite. [*Ham Radio License Manual*, page 6-24]

T8B02   What is the impact of using too much effective radiated power on a satellite uplink?

A. Possibility of commanding the satellite to an improper mode
B. Blocking access by other users
C. Overloading the satellite batteries
D. Possibility of rebooting the satellite control computer

B   Always use the minimum amount of transmitter power to contact satellites, since their relay transmitter power is limited by their solar panels and batteries. If your signal on the satellite downlink is about the same strength as that of the satellite's beacon, you're using the right amount of power. [*Ham Radio License Manual*, page 6-24]

T8B03 Which of the following are provided by satellite tracking programs?

A. Maps showing the real-time position of the satellite track over the earth
B. The time, azimuth, and elevation of the start, maximum altitude, and end of a pass
C. The apparent frequency of the satellite transmission, including effects of Doppler shift
D. All of these choices are correct

D Tracking software provides real-time maps of the satellite's location, the trajectory the satellite will follow across the sky, and even the amount of Doppler shift the signals will experience. You will need to enter data about the satellite's orbit called the *Keplerian elements*. [*Ham Radio License Manual*, page 6-23]

T8B04 What mode of transmission is used by amateur radio satellites?

A. SSB
B. FM
C. CW/data
D. All of these choices are correct

D Satellites can use any amateur signal mode. The most common are SSB, FM, CW, and data. See question T8B08. [*Ham Radio License Manual*, page 6-24]

T8B05   What is a satellite beacon?

A. The primary transmit antenna on the satellite
B. An indicator light that shows where to point your antenna
C. A reflective surface on the satellite
D. A transmission from a satellite that contains status information

D   Common satellite terms include:

• Apogee — The point of a satellite's orbit that is farthest from Earth

• Beacon — A signal from the satellite containing information about a satellite

• Doppler shift — A shift in a signal's frequency due to relative motion between the satellite and the Earth station

• Elliptical orbit — An orbit with a large difference between apogee and perigee

• LEO — A satellite in Low Earth Orbit

• Perigee — The point of a satellite's orbit that is nearest the Earth

• Space station — Defined by the FCC as an amateur station located more than 50 km above the Earth's surface

• Spin fading — Signal fading caused by rotation of the satellite and its antennas.

[*Ham Radio License Manual*, page 6-22]

T8B06   Which of the following are inputs to a satellite tracking program?

A. The weight of the satellite
B. The Keplerian elements
C. The last observed time of zero Doppler shift
D. All of these choices are correct

B   See question T8B03. [*Ham Radio License Manual*, page 6-23]

T8B07   With regard to satellite communications, what is Doppler shift?

A. A change in the satellite orbit
B. A mode where the satellite receives signals on one band and transmits on another
C. An observed change in signal frequency caused by relative motion between the satellite and the earth station
D. A special digital communications mode for some satellites

C   See question T8B05. [*Ham Radio License Manual*, page 6-23]

T8B08 **What is meant by the statement that a satellite is operating in mode U/V?**

A. The satellite uplink is in the 15 meter band and the downlink is in the 10 meter band

B. The satellite uplink is in the 70 centimeter band and the downlink is in the 2 meter band

C. The satellite operates using ultraviolet frequencies

D. The satellite frequencies are usually variable

B A satellite's mode specifies the bands on which it is transmitting and receiving. Mode is specified as two letters separated by a slash. The first letter indicates the uplink band and the second letter indicates the downlink band. For example, the uplink for a satellite in U/V mode is in the UHF band (70 cm) and a downlink is in the VHF band (2 meters). [*Ham Radio License Manual*, page 6-24]

T8B09 **What causes spin fading of satellite signals?**

A. Circular polarized noise interference radiated from the sun

B. Rotation of the satellite and its antennas

C. Doppler shift of the received signal

D. Interfering signals within the satellite uplink band

B See question T8B05. [*Ham Radio License Manual*, page 6-23]

T8B10 **What do the initials LEO tell you about an amateur satellite?**

A. The satellite battery is in Low Energy Operation mode

B. The satellite is performing a Lunar Ejection Orbit maneuver

C. The satellite is in a Low Earth Orbit

D. The satellite uses Light Emitting Optics

C See question T8B05. [*Ham Radio License Manual*, page 6-23]

T8B11 **Who may receive telemetry from a space station?**

A. Anyone who can receive the telemetry signal

B. A licensed radio amateur with a transmitter equipped for interrogating the satellite

C. A licensed radio amateur who has been certified by the protocol developer

D. A licensed radio amateur who has registered for an access code from AMSAT

A Anyone can receive the stream of telemetry data from a space station, even if they don't have an amateur license. [*Ham Radio License Manual*, page 6-24]

T8B12 Which of the following is a good way to judge whether your uplink power is neither too low nor too high?

A. Check your signal strength report in the telemetry data
B. Listen for distortion on your downlink signal
C. Your signal strength on the downlink should be about the same as the beacon
D. All of these choices are correct

C   See question T8B02. [*Ham Radio License Manual*, page 6-24]

T8C — Operating activities: radio direction finding; radio control; contests; linking over the internet; grid locators

T8C01 Which of the following methods is used to locate sources of noise interference or jamming?

A. Echolocation
B. Doppler radar
C. Radio direction finding
D. Phase locking

C   Locating a hidden transmitter (the fox) or *foxhunting* has been a popular ham activity for many years. It has its practical side, too, training hams to find downed aircraft, lost hikers, and sources of interference or jamming. It doesn't require much in the way of equipment. You can get started with a portable radio with a signal strength indicator and a handheld or portable directional antenna, such as a small Yagi beam. [*Ham Radio License Manual*, page 6-10]

T8C02 Which of these items would be useful for a hidden transmitter hunt?

A. Calibrated SWR meter
B. A directional antenna
C. A calibrated noise bridge
D. All of these choices are correct

B   See question T8C01. [*Ham Radio License Manual*, page 6-10]

T8C03 What operating activity involves contacting as many stations as possible during a specified period?

A. Contesting
B. Net operations
C. Public service events
D. Simulated emergency exercises

A   Radio *contests* are held in which the competitors try to make as many short contacts as possible in a fixed period of time. [*Ham Radio License Manual*, page 6-9]

T8C04 Which of the following is good procedure when contacting another station in a radio contest?

A. Sign only the last two letters of your call if there are many other stations calling
B. Contact the station twice to be sure that you are in his log
C. Send only the minimum information needed for proper identification and the contest exchange
D. All of these choices are correct

C When making a contest contact, send only the minimum information needed to identify your station and send the complete information required, called the *exchange*. [*Ham Radio License Manual*, page 6-9]

T8C05 What is a grid locator?

A. A letter-number designator assigned to a geographic location
B. A letter-number designator assigned to an azimuth and elevation
C. An instrument for neutralizing a final amplifier
D. An instrument for radio direction finding

A The Maidenhead Locator System is named for the town outside London, England where the method was first created. In this system, the Earth's surface is divided into a system of rectangles based on latitude and longitude known as *grid locators* or *grid squares*. Each grid square is identified with a combination of letters and numbers. [*Ham Radio License Manual*, page 6-8]

T8C06 How is access to some IRLP nodes accomplished?

A. By obtaining a password that is sent via voice to the node
B. By using DTMF signals
C. By entering the proper internet password
D. By using CTCSS tone codes

B To initiate a contact or exercise a control function on an IRLP-linked repeater, a control code is used. The code is a sequence of DTMF (Dual-tone Multi-Frequency) tones, like dialing a phone number. [*Ham Radio License Manual*, page 6-15]

T8C07   **What is meant by Voice Over Internet Protocol (VoIP) as used in amateur radio?**

A. A set of rules specifying how to identify your station when linked over the internet to another station
B. A set of guidelines for contacting DX stations during contests using internet access
C. A technique for measuring the modulation quality of a transmitter using remote sites monitored via the internet
D. A method of delivering voice communications over the internet using digital techniques

D   VoIP is a type of digital protocol used to deliver voice and audio over the internet. It is used by the IRLP and EchoLink repeater systems. [*Ham Radio License Manual*, page 6-14]

T8C08   **What is the Internet Radio Linking Project (IRLP)?**

A. A technique to connect amateur radio systems, such as repeaters, via the internet using Voice Over Internet Protocol (VoIP)
B. A system for providing access to websites via amateur radio
C. A system for informing amateurs in real time of the frequency of active DX stations
D. A technique for measuring signal strength of an amateur transmitter via the internet

A   The IRLP and EchoLink systems use VoIP (Voice over Internet Protocol) technology to link repeaters. [*Ham Radio License Manual*, page 6-14]

T8C09   **How might you obtain a list of active nodes that use VoIP?**

A. By subscribing to an online service
B. From on line repeater lists maintained by the local repeater frequency coordinator
C. From a repeater directory
D. All of these choices are correct

D   Each repeater in a VoIP repeater network is called a *node*. Nodes are listed in repeater directories, in online lists, and on the internet. [*Ham Radio License Manual*, page 6-14]

T8C10 **What must be done before you may use the EchoLink system to communicate using a repeater?**

A. You must complete the required EchoLink training
B. You must have purchased a license to use the EchoLink software
C. You must be sponsored by a current EchoLink user
D. You must register your call sign and provide proof of license

D You must be a licensed amateur to use EchoLink repeaters. EchoLink allows audio to come from a PC and microphone, so a radio is not necessary but hams are required to send a copy of their license to the EchoLink system administrators to be authorized to use the system. [*Ham Radio License Manual*, page 6-15]

T8C11 **What name is given to an amateur radio station that is used to connect other amateur stations to the internet?**

A. A gateway
B. A repeater
C. A digipeater
D. A beacon

A A *gateway* is a special kind of digital station that provides a connection to the internet via Amateur Radio. Most gateways are set up to *forward* messages. The most common examples are APRS gateways and the Winlink RMS stations. [*Ham Radio License Manual*, page 5-15]

T8D — Non-voice and digital communications: image signals; digital modes; CW; packet radio; PSK31; APRS; error detection and correction; NTSC; amateur radio networking; Digital Mobile/Migration Radio

T8D01 **Which of the following is a digital communications mode?**

A. Packet radio
B. IEEE 802.11
C. JT65
D. All of these choices are correct

D On VHF/UHF, the popular digital modes include:

• Packet radio based on the AX.25 protocol

• B2F protocol for Winlink system messaging

• JT65 for moonbounce and MSK144 for scatter paths

• IEEE 802.11 (WiFi) adapted to amateur use on microwave bands and known as *Broadband-Hamnet* or *high-speed multimedia* (HSMM)

[*Ham Radio License Manual*, page 5-12]

T8D02   What does the term APRS mean?

    A.   Automatic Packet Reporting System
    B.   Associated Public Radio Station
    C.   Auto Planning Radio Set-up
    D.   Advanced Polar Radio System

A   The *Automatic Packet Reporting System* (APRS) uses packet radio to transmit the position information from a moving or portable station. An APRS station is basically a packet radio station combined with a *Global Positioning System* (GPS) receiver. [*Ham Radio License Manual*, page 5-13]

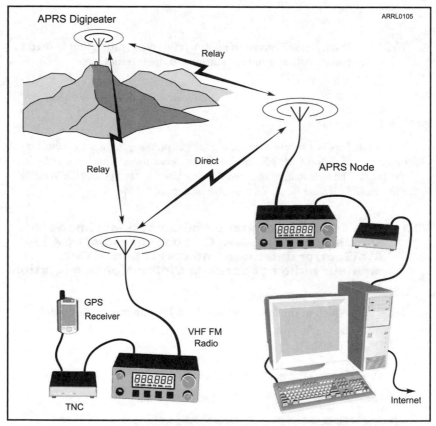

To use the Automatic Packet Reporting System (APRS), the output of a GPS receiver is connected to a packet radio TNC and a VHF radio. Position information and the call sign of the reporting station are then transferred to the APRS system by an APRS node, either directly or via a digipeater. Station location can then be viewed via the Internet.

T8D03 Which of the following devices is used to provide data to the transmitter when sending automatic position reports from a mobile amateur radio station?

A. The vehicle speedometer
B. A WWV receiver
C. A connection to a broadcast FM sub-carrier receiver
D. A Global Positioning System receiver

D See question T8D02. [*Ham Radio License Manual*, page 5-13]

T8D04 What type of transmission is indicated by the term NTSC?

A. A Normal Transmission mode in Static Circuit
B. A special mode for earth satellite uplink
C. An analog fast scan color TV signal
D. A frame compression scheme for TV signals

C NTSC refers to a type of television signal. NTSC (National Television System Committee) fast-scan color television signals are the same as were used for analog broadcast TV signals. [*Ham Radio License Manual*, page 6-10]

T8D05 Which of the following is an application of APRS (Automatic Packet Reporting System)?

A. Providing real-time tactical digital communications in conjunction with a map showing the locations of stations
B. Showing automatically the number of packets transmitted via PACTOR during a specific time interval
C. Providing voice over internet connection between repeaters
D. Providing information on the number of stations signed into a repeater

A Digipeaters and gateways forward the APRS packets, position information, and call sign to a system of server computers via the Internet. Once the information is stored on the servers, websites access the data and show the position of the station on maps in various ways. A common public service application of APRS is to provide maps of station locations while they are providing real-time tactical communications. [*Ham Radio License Manual*, page 5-14]

T8D06 What does the abbreviation PSK mean?

A. Pulse Shift Keying
B. Phase Shift Keying
C. Packet Short Keying
D. Phased Slide Keying

B The most popular keyboard-to-keyboard mode today is PSK31, which stands for *phase shift keying, 31 baud*. [*Ham Radio License Manual*, page 5-13]

T8D07 **Which of the following best describes DMR (Digital Mobile Radio)?**

A. A technique for time-multiplexing two digital voice signals on a single 12.5 kHz repeater channel
B. An automatic position tracking mode for FM mobiles communicating through repeaters
C. An automatic computer logging technique for hands-off logging when communicating while operating a vehicle
D. A digital technique for transmitting on two repeater inputs simultaneously for automatic error correction

A The DMR system was developed for the Land Mobile Radio service. Over the air, DMR is a technique for time-multiplexing two digital voice signals on a single 12.5 kHz repeater channel. [*Ham Radio License Manual*, page 6-15]

T8D08 **Which of the following may be included in packet transmissions?**

A. A check sum that permits error detection
B. A header that contains the call sign of the station to which the information is being sent
C. Automatic repeat request in case of error
D. All of these choices are correct

D Each packet consists of a *header* and *data*. The header contains information about the packet and the call sign of the destination station. The header also includes a *checksum* that allows the receiver to detect errors. If an error is detected, the receiver automatically requests that the packet be retransmitted until the data is received properly. This is called ARQ for *automatic repeat request*. [*Ham Radio License Manual*, page 5-13]

T8D09 **What code is used when sending CW in the amateur bands?**

A. Baudot
B. Hamming
C. International Morse
D. All of these choices are correct

C International Morse is the standard form of code for amateur CW operation. [*Ham Radio License Manual*, page 5-11]

T8D10 Which of the following operating activities is supported by digital mode software in the WSJT suite?

A. Moonbounce or Earth-Moon-Earth
B. Weak-signal propagation beacons
C. Meteor scatter
D. All of these choices are correct

D Many popular digital modes are part of the *WSJT Suite*, a package of open-source software initially developed by Joe Taylor, K1JT. A team maintains and extends the software today, including modes designed for special types of communication such as JT65 for moonbounce (or Earth-Moon-Earth, EME), weak-signal propagation beacons (WSPR), and meteor scatter (MSK144). The latest invention, FT8, is capable of operating in low signal-to-noise conditions by transmitting special code sequences on 15-second intervals. [*Ham Radio License Manual*, page 5-12]

T8D11 What is an ARQ transmission system?

A. A special transmission format limited to video signals
B. A system used to encrypt command signals to an amateur radio satellite
C. A digital scheme whereby the receiving station detects errors and sends a request to the sending station to retransmit the information
D. A method of compressing the data in a message so more information can be sent in a shorter time

C See question T8D08. [*Ham Radio License Manual*, page 5-13]

T8D12 Which of the following best describes Broadband-Hamnet™, also referred to as a high-speed multi-media network?

A. An amateur-radio-based data network using commercial Wi-Fi gear with modified firmware
B. A wide-bandwidth digital voice mode employing DRM protocols
C. A satellite communications network using modified commercial satellite TV hardware
D. An internet linking protocol used to network repeaters

A See question T8D01. [*Ham Radio License Manual*, page 5-12]

T8D13 What is FT8?

A. A wideband FM voice mode
B. A digital mode capable of operating in low signal-to-noise conditions that transmits on 15-second intervals
C. An eight-channel multiplex mode for FM repeaters
D. A digital slow scan TV mode with forward error correction and automatic color compensation

B See question T8D10. [*Ham Radio License Manual*, page 5-12]

T8D14   What is an electronic keyer?

   A. A device for switching antennas from transmit to receive
   B. A device for voice activated switching from receive to transmit
   C. A device that assists in manual sending of Morse code
   D. An interlock to prevent unauthorized use of a radio

C An *electronic keyer* turns contact closures by a Morse *paddle* into a stream of dots and dashes. A keyer may be a standalone device or it can be built into a transceiver. [*Ham Radio License Manual*, page 5-7]

Antennas and Feed Lines

SUBELEMENT T9 — Antennas and feed lines
[2 exam questions — 2 groups]

T9A — Antennas: vertical and horizontal polarization; concept of gain; common portable and mobile antennas; relationships between resonant length and frequency; concept of dipole antennas

T9A01   What is a beam antenna?

    A. An antenna built from aluminum I-beams
    B. An omnidirectional antenna invented by Clarence Beam
    C. An antenna that concentrates signals in one direction
    D. An antenna that reverses the phase of received signals

C   Beam antennas use multiple elements or reflecting surfaces to focus the transmitted signal or receiving ability in a specific direction. [*Ham Radio License Manual*, page 4-16]

T9A02   Which of the following describes a type of antenna loading?

    A. Inserting an inductor in the radiating portion of the antenna to make it electrically longer
    B. Inserting a resistor in the radiating portion of the antenna to make it resonant
    C. Installing a spring in the base of a mobile vertical antenna to make it more flexible
    D. Strengthening the radiating elements of a beam antenna to better resist wind damage

A   Inserting an inductor into an antenna element is called *inductive loading* and it makes the antenna longer electrically than it is physically. [*Ham Radio License Manual*, page 4-13]

T9A03 Which of the following describes a simple dipole oriented parallel to the Earth's surface?

    A. A ground-wave antenna
    B. A horizontally polarized antenna
    C. A rhombic antenna
    D. A vertically polarized antenna

B A dipole radiates a signal with a polarization that is the same as that of the orientation of the dipole. [*Ham Radio License Manual*, page 4-12]

T9A04 What is a disadvantage of the rubber duck antenna supplied with most handheld radio transceivers when compared to a full-sized quarter-wave antenna?

    A. It does not transmit or receive as effectively
    B. It transmits only circularly polarized signals
    C. If the rubber end cap is lost, it will unravel very easily
    D. All of these choices are correct

A The flexible antenna used with most handheld radios is called a *rubber duck*. It's a ground-plane antenna shortened by coiling the conductor inside a plastic coating. The body of the radio and the operator form the antenna's ground plane. The rubber duck is conveniently sized, but doesn't transmit or receive as well as a full-sized ground-plane antenna. [*Ham Radio License Manual*, page 4-15]

T9A05 How would you change a dipole antenna to make it resonant on a higher frequency?

    A. Lengthen it
    B. Insert coils in series with radiating wires
    C. Shorten it
    D. Add capacitive loading to the ends of the radiating wires

C Use an SWR meter or antenna analyzer to determine the resonant frequency. If the resonant frequency is too low, the dipole is too long: shorten it until it is resonant at the desired frequency. If the resonant frequency is too high, you will have to lengthen the dipole. [*Ham Radio License Manual*, page 4-13]

T9A06 What type of antennas are the quad, Yagi, and dish?

    A. Non-resonant antennas
    B. Log periodic antennas
    C. Directional antennas
    D. Isotropic antennas

C Quads, Yagis, and dishes are all examples of *directional* or *beam antennas*. [*Ham Radio License Manual*, page 4-16]

T9A07   What is a disadvantage of using a handheld VHF transceiver, with its integral antenna, inside a vehicle?

A. Signals might not propagate well due to the shielding effect of the vehicle
B. It might cause the transceiver to overheat
C. The SWR might decrease, decreasing the signal strength
D. All of these choices are correct

A   When using a handheld transceiver inside a vehicle, the standard flexible "rubber duck" antenna may not be an effective antenna. The vehicle's metal roof and doors act like shields, trapping the radio waves inside. Some of the signal gets out through the windows (unless they're tinted with a thin metal coating), but it's as much as 10 to 20 times weaker than an external mobile antenna. [*Ham Radio License Manual*, page 4-15]

T9A08   What is the approximate length, in inches, of a quarter-wavelength vertical antenna for 146 MHz?

A. 112
B. 50
C. 19
D. 12

C   The length of a ground-plane antenna is half that of a dipole so use the formula:

Length (in feet) = 234 / frequency (in MHz)

At 146 MHz, a $\lambda/4$ (¼ λ) ground-plane is 234 / 146 = 1.6 feet = 19¼ inches long.

[*Ham Radio License Manual*, page 4-13]

T9A09   What is the approximate length, in inches, of a half-wavelength 6 meter dipole antenna?

A. 6
B. 50
C. 112
D. 236

C   Use the formula:

Length (in feet) = 468 / frequency (in MHz)

At 50.1 MHz (in the 6 meter band), dipole length is calculated as 468 / 50.1 = 9.33 feet = 112 inches long

[*Ham Radio License Manual*, page 4-13]

T9A10 **In which direction does a half-wave dipole antenna radiate the strongest signal?**

A. Equally in all directions
B. Off the ends of the antenna
C. Broadside to the antenna
D. In the direction of the feed line

C A dipole radiates strongest broadside to the antenna and weakest off the ends. [*Ham Radio License Manual*, page 4-12]

T9A11 **What is the gain of an antenna?**

A. The additional power that is added to the transmitter power
B. The additional power that is lost in the antenna when transmitting on a higher frequency
C. The increase in signal strength in a specified direction compared to a reference antenna
D. The increase in impedance on receive or transmit compared to a reference antenna

C Concentrating an antenna's radiated signals in a specific direction is called *gain*. Antenna gain increases signal strength in a specified direction when compared to a reference antenna. [*Ham Radio License Manual*, page 4-7]

T9A12 **What is an advantage of using a properly mounted 5/8 wavelength antenna for VHF or UHF mobile service?**

A. It has a lower radiation angle and more gain than a ¼ wavelength antenna
B. It has very high angle radiation for better communicating through a repeater
C. It eliminates distortion caused by reflected signals
D. It has 10 times the power gain of a ¼ wavelength design

A Due to its extended length compared to a ¼-λ antenna, the ⅝-λ antenna focuses a bit more energy toward the horizon, improving range. [*Ham Radio License Manual,* page 4-12]

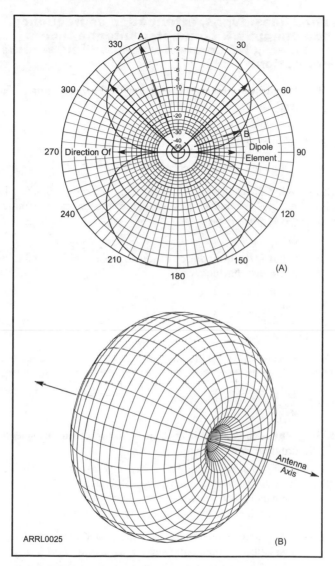

The radiation pattern of a dipole far from ground (in free-space). At (A) the pattern is shown in a plane containing the dipole. The lengths of the arrows indicate the relative strength of the radiated power in that direction. The dipole radiates best broadside to its length. At (B) the 3-D pattern shows radiated strength in all directions.

T9B — Feed lines: types, attenuation vs frequency, selecting; SWR concepts; Antenna tuners (couplers); RF Connectors: selecting, weather protection

T9B01   Why is it important to have low SWR when using coaxial cable feed line?

A. To reduce television interference
B. To reduce signal loss
C. To prolong antenna life
D. All of these choices are correct

B   Low SWR reduces losses in the feed line from reflected power in the feed line traveling back and forth between the antenna and transmitter. [*Ham Radio License Manual*, page 4-11]

T9B02   What is the impedance of most coaxial cables used in amateur radio installations?

A. 8 ohms
B. 50 ohms
C. 600 ohms
D. 12 ohms

B   Most coaxial cable used in ham radio has a characteristic impedance (Z_0) of 50 ohms. Coaxial cables used for video and cable television have a Z_0 of 75 ohms. Open-wire feed lines have a Z_0 of 300 to 600 ohms. [*Ham Radio License Manual*, page 4-9]

T9B03   Why is coaxial cable the most common feed line selected for amateur radio antenna systems?

A. It is easy to use and requires few special installation considerations
B. It has less loss than any other type of feed line
C. It can handle more power than any other type of feed line
D. It is less expensive than any other type of feed line

A   "Coax" is flexible, is unaffected by weather, and can be routed in bundles and next to insulating or conducting supports. [*Ham Radio License Manual*, page 4-9]

T9B04 **What is the major function of an antenna tuner (antenna coupler)?**

A. It matches the antenna system impedance to the transceiver's output impedance
B. It helps a receiver automatically tune in weak stations
C. It allows an antenna to be used on both transmit and receive
D. It automatically selects the proper antenna for the frequency band being used

A An antenna tuner is used to transform the antenna system's impedance to match that of the transceiver's output amplifier. [*Ham Radio License Manual*, page 4-19]

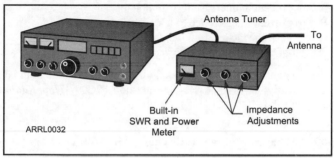

An antenna tuner acts like an electrical version of a mechanical gearbox. By adjusting the tuner's controls, the impedance present at the end of the feed line can be converted to the impedance that best suits the transceiver's output circuits, usually 50 Ω.

T9B05 **In general, what happens as the frequency of a signal passing through coaxial cable is increased?**

A. The characteristic impedance decreases
B. The loss decreases
C. The characteristic impedance increases
D. The loss increases

D Feed line loss increases with frequency for all types of feed lines. [*Ham Radio License Manual*, page 4-9]

T9B06 **Which of the following connectors is most suitable for frequencies above 400 MHz?**

A. A UHF (PL-259/SO-239) connector
B. A Type N connector
C. An RS-213 connector
D. A DB-25 connector

B Above 400 MHz, the Type N connectors are used because of their low loss and controlled impedance. [*Ham Radio License Manual*, page 4-18]

T9B07   Which of the following is true of PL-259 type coax connectors?

A. They are preferred for microwave operation
B. They are watertight
C. They are commonly used at HF frequencies
D. They are a bayonet type connector

C   The UHF series of connectors — PL-259 plugs and SO-239 receptacles — are the most widely-used for HF equipment. [*Ham Radio License Manual*, page 4-18]

T9B08   Why should coax connectors exposed to the weather be sealed against water intrusion?

A. To prevent an increase in feed line loss
B. To prevent interference to telephones
C. To keep the jacket from becoming loose
D. All of these choices are correct

A   Water in coaxial cable degrades the effectiveness of the braided shield and dramatically increases losses. See question T7C09. [*Ham Radio License Manual*, page 4-18]

T9B09   What can cause erratic changes in SWR readings?

A. The transmitter is being modulated
B. A loose connection in an antenna or a feed line
C. The transmitter is being over-modulated
D. Interference from other stations is distorting your signal

B   An intermittent connection creates large mismatches in impedance. This causes large erratic changes in SWR. [*Ham Radio License Manual*, page 4-11]

T9B10   What is the electrical difference between RG-58 and RG-8 coaxial cable?

A. There is no significant difference between the two types
B. RG-58 cable has two shields
C. RG-8 cable has less loss at a given frequency
D. RG-58 cable can handle higher power levels

C   In general, a larger diameter cable such as RG-8 will have less loss than a small cable such as RG-58. [*Ham Radio License Manual*, page 4-17]

T9B11 Which of the following types of feed line has the lowest loss at VHF and UHF?

A. 50-ohm flexible coax
B. Multi-conductor unbalanced cable
C. Air-insulated hard line
D. 75-ohm flexible coax

C A special type of coaxial feed line is called *hardline* because its shield is made from a semi-flexible solid tube of aluminum or copper. It has the lowest loss of any type of coaxial feed line. [*Ham Radio License Manual*, page 4-9]

Electrical Safety

SUBELEMENT T0 — Electrical safety: AC and DC power circuits; antenna installation; RF hazards
[3 exam questions — 3 groups]

T0A — Power circuits and hazards: hazardous voltages; fuses and circuit breakers; grounding; lightning protection; battery safety; electrical code compliance

T0A01 Which of the following is a safety hazard of a 12-volt storage battery?

A. Touching both terminals with the hands can cause electrical shock
B. Shorting the terminals can cause burns, fire, or an explosion
C. RF emissions from the battery
D. All of these choices are correct

B Storage batteries release a lot of energy if shorted, leading to burns, fire, or an explosion. Keep metal objects such as tools and sheet metal clear of battery terminals and avoid working on equipment with the battery connected. [*Ham Radio License Manual*, page 9-3]

T0A02 What health hazard is presented by electrical current flowing through the body?

A. It may cause injury by heating tissue
B. It may disrupt the electrical functions of cells
C. It may cause involuntary muscle contractions
D. All of these choices are correct

D The severity of the hazard varies with the amount of current as shown in the table on the next page. [*Ham Radio License Manual*, page 9-2]

Effects of Electric Current in the Human Body

Current	Reaction
Below 1 milliampere	Generally not perceptible
1 milliampere	Faint tingle
5 milliamperes	Slight shock felt; not painful but disturbing. Average individual can let go. Strong involuntary reactions can lead to other injuries.
6-25 milliamperes (women) 9-30 milliamperes (men)	Painful shock, loss of muscular control; the freezing current or "can't let-go" range.
50-150 milliamperes	Extreme pain, respiratory arrest, severe muscular contractions. Death is possible.
1000-4300 milliamperes	Rhythmic pumping action of the heart ceases. Muscular contraction and nerve damage occur; death likely.
10,000 milliamperes	Cardiac arrest, severe burns; death probable.

T0A03   In the United States, what is connected to the green wire in a three-wire electrical AC plug?

A. Neutral
B. Hot
C. Equipment ground
D. The white wire

C   The US standard is hot – black wire (occasionally red); neutral – white wire; and safety or equipment ground – green or bare wire. [*Ham Radio License Manual*, page 9-5]

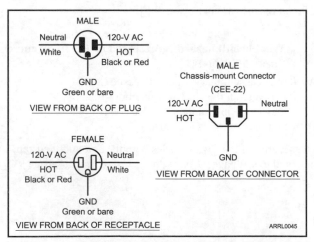

The correct wiring technique for 120 V ac power cords and receptacles. The white wire is neutral and the green wire is the safety ground. The hot wire can be either black or red. These receptacles are shown from the back, or wiring side.

T0A04 **What is the purpose of a fuse in an electrical circuit?**

A. To prevent power supply ripple from damaging a circuit
B. To interrupt power in case of overload
C. To limit current to prevent shocks
D. All of these choices are correct

B   When the metal in a fuse melts or "blows," the current path is broken and power is removed from circuits supplied by the fuse. [*Ham Radio License Manual*, page 3-12]

T0A05 **Why is it unwise to install a 20-ampere fuse in the place of a 5-ampere fuse?**

A. The larger fuse would be likely to blow because it is rated for higher current
B. The power supply ripple would greatly increase
C. Excessive current could cause a fire
D. All of these choices are correct

C   Replacing a blown fuse with one having a higher current rating, even temporarily, could allow the fault to permanently damage the equipment or start a fire. Do not use a device with a higher current rating, even temporarily. [*Ham Radio License Manual*, page 3-12]

T0A06 **What is a good way to guard against electrical shock at your station?**

A. Use three-wire cords and plugs for all AC powered equipment
B. Connect all AC powered station equipment to a common safety ground
C. Use a circuit protected by a ground-fault interrupter
D. All of these choices are correct

D   Follow these simple rules in your station:

• Use three-wire power cords and plugs for all ac-powered equipment.

• Make sure all of your equipment has a connection to the ac safety ground.

• Use *ground fault circuit interrupter (GFCI)* circuit breakers or circuit breaker outlets.

• Verify ac wiring is done properly by using an ac circuit tester.

• Never replace a fuse or circuit breaker with one of a larger size.

• Don't overload single outlets.

[*Ham Radio License Manual*, page 9-4]

T0A07 Which of these precautions should be taken when installing devices for lightning protection in a coaxial cable feed line?

A. Include a parallel bypass switch for each protector so that it can be switched out of the circuit when running high power

B. Include a series switch in the ground line of each protector to prevent RF overload from inadvertently damaging the protector

C. Keep the ground wires from each protector separate and connected to station ground

D. Mount all of the protectors on a metal plate that is in turn connected to an external ground rod

D See question T0B10. [*Ham Radio License Manual*, page 9-5]

T0A08 What safety equipment should always be included in home-built equipment that is powered from 120V AC power circuits?

A. A fuse or circuit breaker in series with the AC hot conductor

B. An AC voltmeter across the incoming power source

C. An inductor in parallel with the AC power source

D. A capacitor in series with the AC power source

A Fuses or circuit breakers must be installed in series with the hot conductor or conductors so that if activated or *tripped*, power is removed from the protected equipment. [*Ham Radio License Manual*, page 9-5]

T0A09 What should be done to all external ground rods or earth connections?

A. Waterproof them with silicone caulk or electrical tape

B. Keep them as far apart as possible

C. Bond them together with heavy wire or conductive strap

D. Tune them for resonance on the lowest frequency of operation

C See question T0B10. [*Ham Radio License Manual*, page 9-5]

T0A10 What can happen if a lead-acid storage battery is charged or discharged too quickly?

A. The battery could overheat, give off flammable gas, or explode

B. The voltage can become reversed

C. The memory effect will reduce the capacity of the battery

D. All of these choices are correct

A Storage batteries hold a lot of energy and must be treated with respect. They contain strong acids that can be hazardous if spilled or allowed to leak. Storage batteries can also release or vent flammable hydrogen gas, and that can cause an explosion. [*Ham Radio License Manual*, page 10-19]

T0A11 What kind of hazard might exist in a power supply when it is turned off and disconnected?

A. Static electricity could damage the grounding system
B. Circulating currents inside the transformer might cause damage
C. The fuse might blow if you remove the cover
D. You might receive an electric shock from the charge stored in large capacitors

D Capacitors in a power supply can store charge after a charging circuit is turned off, presenting a hazardous voltage for a long time. This includes small-value capacitors charged to a high voltage. Make sure capacitors are discharged by testing them with a meter or use a *grounding stick* to shunt their charge to ground. [*Ham Radio License Manual*, page 9-2]

T0B — Antenna safety: tower safety and grounding; erecting an antenna support; safely installing an antenna

T0B01 When should members of a tower work team wear a hard hat and safety glasses?

A. At all times except when climbing the tower
B. At all times except when belted firmly to the tower
C. At all times when any work is being done on the tower
D. Only when the tower exceeds 30 feet in height

C Starting with personal preparation, both climbers and ground crew should wear appropriate protective gear any time work is under way on the tower. Each member of the crew should wear a hard hat, goggles or safety glasses, and heavy duty gloves suitable for working with ropes. If you are the climber, use an inspected and approved climbing harness (fall arrester) and work boots to protect the arches of your feet. [*Ham Radio License Manual*, page 9-19]

T0B02 What is a good precaution to observe before climbing an antenna tower?

A. Make sure that you wear a grounded wrist strap
B. Remove all tower grounding connections
C. Put on a carefully inspected climbing harness (fall arrester) and safety glasses
D. All of these choices are correct

C See question T0B01. [*Ham Radio License Manual*, page 9-19]

T0B03 Under what circumstances is it safe to climb a tower without a helper or observer?

A. When no electrical work is being performed
B. When no mechanical work is being performed
C. When the work being done is not more than 20 feet above the ground
D. Never

D Having a ground crew is important; avoid climbing alone whenever possible because it's never safe. If you do climb alone, take along a handheld radio. A ground crew should have enough members to do the job safely, including rendering aid if necessary. [*Ham Radio License Manual*, page 9-20]

T0B04 Which of the following is an important safety precaution to observe when putting up an antenna tower?

A. Wear a ground strap connected to your wrist at all times
B. Insulate the base of the tower to avoid lightning strikes
C. Look for and stay clear of any overhead electrical wires
D. All of these choices are correct

C Power lines are the enemy of antenna installers. Place all antennas and feed lines well clear of power lines, including the utility service drop to your home. Be sure that if the any part of the antenna or support structure falls, it cannot fall onto power lines. A good guideline is to separate the antenna from the nearest power line by 150% of total height of tower or mast plus antenna — a minimum of 10 feet of clearance during a fall is a must. Never attach an antenna or guy wire to a utility pole, since a mechanical failure could result in contact with high-voltage power lines. [*Ham Radio License Manual*, page 9-18]

T0B05 What is the purpose of a gin pole?

A. To temporarily replace guy wires
B. To be used in place of a safety harness
C. To lift tower sections or antennas
D. To provide a temporary ground

C Gin poles are temporary masts used to lift materials such as antennas or tower sections so that you do not have to hoist things directly. [*Ham Radio License Manual*, page 9-20]

T0B06 What is the minimum safe distance from a power line to allow when installing an antenna?

A. Half the width of your property
B. The height of the power line above ground
C. 1/2 wavelength at the operating frequency
D. Enough so that if the antenna falls unexpectedly, no part of it can come closer than 10 feet to the power wires

D See question T0B04. [*Ham Radio License Manual*, page 9-18]

T0B07   Which of the following is an important safety rule to remember when using a crank-up tower?

A. This type of tower must never be painted
B. This type of tower must never be grounded
C. This type of tower must not be climbed unless retracted or mechanical safety locking devices have been installed
D. All of these choices are correct

C   Crank-up towers must be fully retracted or mechanical safety locking devices must have been installed. Never climb a crank-up tower supported only by the cable that supports the sections. [*Ham Radio License Manual*, page 9-20]

T0B08   What is considered to be a proper grounding method for a tower?

A. A single four-foot ground rod, driven into the ground no more than 12 inches from the base
B. A ferrite-core RF choke connected between the tower and ground
C. Separate eight-foot long ground rods for each tower leg, bonded to the tower and each other
D. A connection between the tower base and a cold water pipe

C   Grounding rules for antennas and supports must be followed according to your local electrical code. Towers should be grounded with separate 8-foot long ground rods for each tower leg, bonded to the tower and each other. [*Ham Radio License Manual*, page 9-18]

T0B09   Why should you avoid attaching an antenna to a utility pole?

A. The antenna will not work properly because of induced voltages
B. The utility company will charge you an extra monthly fee
C. The antenna could contact high-voltage power lines
D. All of these choices are correct

C   See question T0B04. [*Ham Radio License Manual*, page 9-18]

T0B10 Which of the following is true when installing grounding conductors used for lightning protection?

A. Only non-insulated wire must be used
B. Wires must be carefully routed with precise right-angle bends
C. Sharp bends must be avoided
D. Common grounds must be avoided

C Starting at your antennas, all towers, masts, and antenna mounts should be grounded according to your local building and electrical codes. These connections are made at the tower base, or in the case of roof mounts, though a large-diameter wire to a ground rod. Ground connections should be as short and direct as possible — avoid sharp bends. Where cables and feed lines enter the house, use lightning arrestors grounded to a common plate that is in turn connected to a nearby external ground such as a ground rod. All ground rods and earth connections must be bonded together with heavy wire, as well [*Ham Radio License Manual*, page 9-5]

T0B11 Which of the following establishes grounding requirements for an amateur radio tower or antenna?

A. FCC Part 97 Rules
B. Local electrical codes
C. FAA tower lighting regulations
D. UL recommended practices

B See question T0B10. [*Ham Radio License Manual*, page 9-5]

T0B12 Which of the following is good practice when installing ground wires on a tower for lightning protection?

A. Put a loop in the ground connection to prevent water damage to the ground system
B. Make sure that all bends in the ground wires are clean, right-angle bends
C. Ensure that connections are short and direct
D. All of these choices are correct

C See question T0B10. [*Ham Radio License Manual*, page 9-5]

T0B13 What is the purpose of a safety wire through a turnbuckle used to tension guy lines?

A. Secure the guy if the turnbuckle breaks
B. Prevent loosening of the guy line from vibration
C. Prevent theft or vandalism
D. Deter unauthorized climbing of the tower

B Safety wire through turnbuckles used to tension guy lines prevents them from loosening due to vibration and twisting. [*Ham Radio License Manual*, page 9-18]

T0C — RF hazards: radiation exposure; proximity to antennas; recognized safe power levels; exposure to others; radiation types; duty cycle

T0C01   What type of radiation are VHF and UHF radio signals?

A. Gamma radiation
B. Ionizing radiation
C. Alpha radiation
D. Non-ionizing radiation

D   RF radiation is not the same as *ionizing radiation* from radioactivity because the energy in signals at radio frequencies is far too low to cause an electron to leave an atom (ionize) and therefore cannot cause genetic damage. With its relatively low frequency, RF energy is *non-ionizing radiation*. [*Ham Radio License Manual*, page 9-11]

T0C02   Which of the following frequencies has the lowest value for Maximum Permissible Exposure limit?

A. 3.5 MHz
B. 50 MHz
C. 440 MHz
D. 1296 MHz

B   Using the following graph to compare MPE for amateur bands at 3.5, 50, 440, and 1296 MHz, you can see that MPE is lowest at 50 MHz and highest at 3.5 MHz. [*Ham Radio License Manual*, page 9-12]

Maximum Permissible Exposure (MPE) Limits

Controlled Exposure (6-Minute Average)

Frequency Range (MHz)	Power Density (mW/cm²)
0.3 – 3.0	(100)*
3.0 – 30	(900/f2)*
30 – 300	1.0
300 – 1500	f/300
1500 – 100,000	5

Uncontrolled Exposure (30-Minute Average)

Frequency Range (MHz)	Magnetic Field Power Density (mW/cm²)
0.3 – 1.34	(100)*
1.34 – 30	(180/f2)*
30 – 300	0.2
300 – 1500	f/1500
1500 – 100,000	1.0

f = frequency in MHz
* = Plane-wave equivalent power density

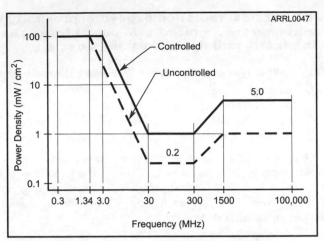

ARRL0047

Maximum Permissible Exposure (MPE) limits vary with frequency because the body responds differently to energy at different frequencies. The controlled and uncontrolled limits refer to the environment in which people are exposed to the RF energy.

T0C03 What is the maximum power level that an amateur radio station may use at VHF frequencies before an RF exposure evaluation is required?

A. 1500 watts PEP transmitter output
B. 1 watt forward power
C. 50 watts PEP at the antenna
D. 50 watts PEP reflected power

C If the transmitter power (using PEP) to the antenna is less than the levels shown in the following table on the frequencies at which you operate, then no evaluation is required. [*Ham Radio License Manual*, page 9-14]

Power Thresholds for RF Exposure Evaluation

Band	Power (W)	Band	Power (W)
160 meters	500	6	50
80	500	2	50
40	500	1.25	50
30	425	70 cm	70
20	225	33	150
17	125	23	200
15	100	13	250
12	75	SHF (all bands)	250
10	50	EHF (all bands)	250

T0C04 **What factors affect the RF exposure of people near an amateur station antenna?**

A. Frequency and power level of the RF field
B. Distance from the antenna to a person
C. Radiation pattern of the antenna
D. All of these choices are correct

D When performing an RF exposure evaluation, you'll need information on the RF signal's frequency and power level, distance from the antenna, and the antenna's radiation pattern. Once you've done an evaluation, you don't need to re-evaluate unless you change equipment in your station that affects average output power, such as increasing transmitter power or antenna gain. You'll also need to re-evaluate if you add a new frequency band. [*Ham Radio License Manual*, page 9-14]

T0C05 **Why do exposure limits vary with frequency?**

A. Lower frequency RF fields have more energy than higher frequency fields
B. Lower frequency RF fields do not penetrate the human body
C. Higher frequency RF fields are transient in nature
D. The human body absorbs more RF energy at some frequencies than at others

D Heating as a result of exposure to RF fields is caused by the body absorbing RF energy. Absorption varies with frequency because the body absorbs more RF energy at some frequencies than others. [*Ham Radio License Manual*, page 9-11]

T0C06 **Which of the following is an acceptable method to determine that your station complies with FCC RF exposure regulations?**

A. By calculation based on FCC OET Bulletin 65
B. By calculation based on computer modeling
C. By measurement of field strength using calibrated equipment
D. All of these choices are correct

D By far the most common evaluation uses the techniques outlined in the FCC's OET Bulletin 65 (OET stands for *Office of Engineering Technology*). This method uses tables and simple formulas to evaluate whether your station has the potential of causing an exposure hazard. [*Ham Radio License Manual*, page 9-14]

T0C07   **What could happen if a person accidentally touched your antenna while you were transmitting?**

A. Touching the antenna could cause television interference
B. They might receive a painful RF burn
C. They might develop radiation poisoning
D. All of these choices are correct

B   RF burns caused by touching or coming close to conducting surfaces with a high RF voltage present are also an effect of heating. While these are sometimes painful, they are rarely hazardous. RF burns can be eliminated by proper bonding techniques or by preventing access to an antenna. [*Ham Radio License Manual*, page 9-11]

T0C08   **Which of the following actions might amateur operators take to prevent exposure to RF radiation in excess of FCC-supplied limits?**

A. Relocate antennas
B. Relocate the transmitter
C. Increase the duty cycle
D. All of these choices are correct

A   Locate antennas away from where people can get close to them and away from property lines. [*Ham Radio License Manual*, page 9-14]

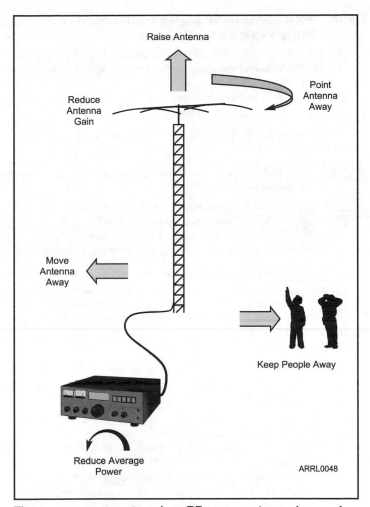

Raise Antenna

Reduce Antenna Gain

Point Antenna Away

Move Antenna Away

Keep People Away

Reduce Average Power

ARRL0048

There are many ways to reduce RF exposure to nearby people. Whatever lowers the power density in areas where people are will work. Raising the antenna will even benefit your signal strength to other stations as it lowers power density on the ground!

T0C09 How can you make sure your station stays in compliance with RF safety regulations?

A. By informing the FCC of any changes made in your station
B. By re-evaluating the station whenever an item of equipment is changed
C. By making sure your antennas have low SWR
D. All of these choices are correct

B See question T0C04. [*Ham Radio License Manual*, page 9-14]

T0C10 Why is duty cycle one of the factors used to determine safe RF radiation exposure levels?

A. It affects the average exposure of people to radiation
B. It affects the peak exposure of people to radiation
C. It takes into account the antenna feed line loss
D. It takes into account the thermal effects of the final amplifier

A Duty cycle is the ratio of the transmitted signal's on-the-air time to the total operating time during the measurement period and has a maximum of 100%. Stated simply, duty cycle is the percentage of time a transmitter is transmitting. Since duty cycle affects the average power level of transmissions, it must be considered when evaluating exposure. [*Ham Radio License Manual*, page 9-13]

T0C11 What is the definition of duty cycle during the averaging time for RF exposure?

A. The difference between the lowest power output and the highest power output of a transmitter
B. The difference between the PEP and average power output of a transmitter
C. The percentage of time that a transmitter is transmitting
D. The percentage of time that a transmitter is not transmitting

C See question T0C10. [*Ham Radio License Manual*, page 9-13]

T0C12 How does RF radiation differ from ionizing radiation (radioactivity)?

A. RF radiation does not have sufficient energy to cause genetic damage
B. RF radiation can only be detected with an RF dosimeter
C. RF radiation is limited in range to a few feet
D. RF radiation is perfectly safe

A See question T0C01. [*Ham Radio License Manual*, page 9-11]

T0C13 If the averaging time for exposure is 6 minutes, how much power density is permitted if the signal is present for 3 minutes and absent for 3 minutes rather than being present for the entire 6 minutes?

A. 3 times as much
B. 1/2 as much
C. 2 times as much
D. There is no adjustment allowed for shorter exposure times

C The lower the duty cycle (less transmitting), the higher the transmitter output can be and still have an average value within the exposure limits. For example, what is the result if a transmitted signal in a controlled environment is present for 3 minutes and then absent for the remaining 3 minutes of the averaging period? Because the signal is only present for 1/2 of the time (50% duty cycle), the signal power can be twice as high and still have the same average power as it would if transmitted continuously with a duty cycle of 100%. [*Ham Radio License Manual*, page 9-13]

Notes

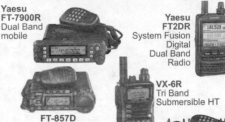

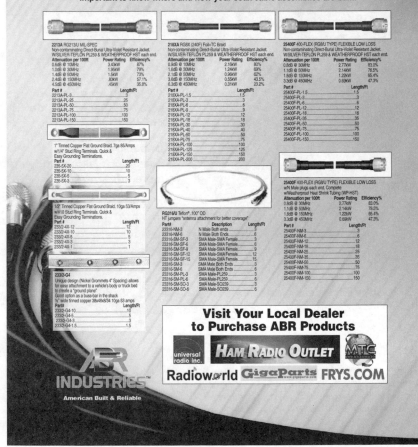